MW00643356

DYNAMICS OF REASON

Stanford Kant Lectures

The Stanford Kant Lectures are an invited series sponsored by the Stanford University department of philosophy. Philosophers are invited to reflect on the state of the art in their field and to report on current research in a way that is accessible to the broader community of scholars, educators, students and the public at large. Topics range from Kant's philosophy of science to the nature of human desire. The Kant Lecturer is selected by the department in light of its judgment that the speaker has a deep understanding of the issues in his or her field, and a vision of their connection to questions of lasting concern across academic disciplines.

MICHAEL FRIEDMAN

Dynamics of Reason

The 1999 Kant Lectures at Stanford University

CSLI Publications
Center for the Study of
Language and Information
Stanford , California

Library of Congress Cataloging-in-Publication Data

Friedman, Michael, 1947–
    Dynamics of reason / Michael Friedman.
        p. cm.
    (Stanford Kant Lectures)
    Includes bibliographical references and index.
    ISBN 1-57586-291-3 (alk. paper)—ISBN 1-57586-292-1 (pbk. : alk. paper)
    1. Science–Philosophy. I. Title

Q175 .F892 2001
501–dc21

2001017171

Copyright © 2001 Michael Friedman

In memory of
Carl Gustav ("Peter") Hempel
teacher, mentor, friend

It might be supposed that [reason's demand for systematic unity and simplicity] is merely an economical contrivance whereby reason seeks to save itself all possible trouble, a hypothetical attempt, which, if it succeeds, will, through the unity thus attained, impart probability to the presumed principle of explanation. But such a selfish purpose can very easily be distinguished from the Idea. For in conformity with the Idea everyone presupposes that this unity of reason accords with nature itself, and that reason – although indeed unable to determine the limits of this unity – does not here beg but command.

—Kant

Contents

Preface

For the past twenty years I have been pursuing historical studies on the relationship between developments in scientific philosophy from Kant (in the late eighteenth century) to logical empiricism (in the early twentieth) and concurrent developments taking place within the sciences, principally within the exact sciences of logic, mathematics, and mathematical physics. These studies concentrated on Kant's involvement with the sciences of his time (Euclidean geometry and Newtonian physics, for example) and the various ways in which late nineteenth and early twentieth century revolutionary changes within these sciences (non-Euclidean geometry and relativistic physics, for example) motivated early twentieth century logical empiricists to undertake similarly revolutionary changes in the Kantian and neo-Kantian philosophy of their time. My primary aim was to depict the deep and intricate connections between the historical development of scientific philosophy, on the one hand, and the parallel evolution of the exact sciences themselves, on the other.

I did not, however, attempt to draw substantive morals for our contemporary philosophical predicament from these historical studies. Indeed, my dominant tone was rather unhopeful. Kant's original philosophical synthesis had failed due to unforeseen revolutionary changes within the sciences, and the logical empiricists's radical revision of this synthesis had also failed to do justice to the very rapid changes taking place within early twentieth century science. In response to this situation I offered only the indeterminate suggestion that we, too, should attempt a fruitful interaction between contemporary philosophical practice and developments within the sciences themselves, and the pessimistic conclusion that we now have no idea how to go about this given the twin failures of Kantianism and logical empiricism.

In the mid to late 1990s I began to see a way out of this dilemma. It was not yet clear how one could preserve any kind of commitment to a Kantian or neo-Kantian conception of a priori principles in the exact sciences (as in Kant's original conception of the synthetic a priori, for example, or Rudolf Carnap's version of the analytic a priori developed in the logical empiricist tradition), but I was convinced, at the same time, that the dominant viewpoint within contemporary scientific philosophy – some or another version of naturalistic epistemological holism – is entirely incapable of providing an adequate philosophical perspective on these sciences. The idea I then came up with, against this twofold background, was that one could attempt to combine basic aspects of Carnap's philosophy of formal languages or linguistic frameworks with fundamental features of Thomas Kuhn's much less formal theory of scientific revolutions. And in this way, more specifically, one could articulate a conception of dynamical or relativized a priori principles within an historical account of the conceptual evolution of the sciences rather than a purely syntactic or semantic account of the formal language(s) of the sciences. One could thus reap the benefits of Kuhnian conceptual history while simultaneously avoiding the drawbacks of Carnapian formal *Wissenschaftslogik*. In particular, W. V. Quine's well-known and widely accepted attack on the Carnapian conception of analytic truth need no longer compel us to adopt a thoroughgoing epistemological holism according to which there is nothing left of the a priori at all.

In 1997 I had the opportunity to deliver the Whitehead Lecture at Harvard University and the Presidential Address to the Central Division of the American Philosophical Association. I chose, for both occasions, the topic of "Philosophical Naturalism" [Friedman (1997a)], where my aim, in particular, was to attack the contemporary consensus in favor of Quinean epistemological holism on behalf of a neo-Kantian (and historicized) conception of relativized a priori principles. I found, at the same time, that I was thereby able to develop a new positive conception of the peculiar role of scientific philosophy vis-à-vis the sciences, as a distinctive meta-scientific level of thought or discourse making its own indispensable contribution to the rationality of radical conceptual revolutions (Kuhnian "paradigm-shifts") taking place within the sciences themselves. I further developed these ideas in a second paper, "On the Idea of a Scientific Philosophy," which I then delivered as part of a series on "Science and Defining the Human" at Northwestern University, as the Stice Lecture at the University of Washington, as a Distinguished Visitor at the University of California at Riverside, as the Jacobson Lecture at the University of London, the Hall Lecture at the University of Iowa, and the Werkmeister Lecture at Florida State University.

This last paper then became the first lecture in the three-lecture series that comprises Part One of the present volume. In the academic year 1998–99 I had the opportunity to deliver this three-lecture series on four different occasions: as the Simon Lectures at the University of Toronto, as part of my annual visit to the University of Western Ontario, as the Hempel Lectures at Princeton University, and as the Kant Lectures at Stanford University. I was especially pleased to have had the opportunity to present the Hempel Lectures at Princeton, both because I had been a student of Hempel's at Princeton in the early 1970s and because my attempt at combining Kuhnian ideas with aspects of logical empiricism was complementary, in a way, to Hempel's own efforts to undertake a parallel accommodation in the later part of his philosophical career – beginning in the early 1970s and extending into the early 1990s [see, in particular, the essays from this period collected in Hempel (2000)]. Whereas Hempel had focussed on the kinship between the pragmatic and anti-formalistic dimensions of Kuhn's thought and Otto Neurath's naturalistic version of logical empiricism (by which Hempel himself had been significantly influenced during his early years in Berlin and Vienna), I was attempting to exploit the kinship between the Kuhnian notion of scientific paradigm and the Carnapian notion of linguistic framework so as to move scientific epistemology in a less pragmatically naturalistic and more Kantian direction.

Finally, I had the opportunity to discuss the ideas of the three lectures at the Fifth Göttinger Philosophisches Kolloquium at the University of Göttingen in June of 1999. In this colloquium series students in the Göttingen philosophical seminar study the work in progress of a particular philosopher during the academic year, followed by a three-day session at the end where papers on this work are presented by the students and intensively discussed with the philosopher in question. On this occasion I had provided the three lectures, together with related materials such as Friedman (1997a), (1997b), (1998), and (2000a). I am deeply indebted to the organizers of the colloquium, Felix Mühlhölzer and Olaf Müller, and, most especially, to the students who prepared papers on my work: Marita Hübner, Lorenz Kähler, Tobias Klauk, Christoph Martel, Uwe Rose, Thomas Schmidt, Carsten Schmidt-Samoa, Frank Tschepke, and Regina Wenninger. I am particularly grateful for this kind of critical feedback, for the ideas presented in the lectures very much represent work in progress and have by no means gelled into final form.

It is for this reason, above all, that I here offer the lectures together with a Part Two containing "Fruits of Discussion." Whereas the lectures themselves are presented basically unchanged (except for footnotes providing references and indications where my thinking has further developed in response to discussion), Part Two attempts a new systematic development of the ideas of the lectures, motivated, in large part, by the many helpful comments I received at all the occasions noted above. At Harvard I am especially indebted to comments from Stanley Cavell, Warren Goldfarb, Christine Korsgaard, Tim Maudlin, Hilary Putnam, and Thomas Scanlon; at the University of Toronto to James Robert Brown and Ian Hacking; at the University of Western Ontario to William Demopoulos, Robert DiSalle, William Harper, Wayne Myrvold, John Nicholas, and Itamar Pitowsky; at Princeton to Frank Arntzenius, Gordon Belot, Paul Benacerraf, John Burgess, Richard Jeffrey, Mark Johnston, Gideon Rosen, and Bas van Fraassen; at Stanford to Michael Bratman, Erik Curiel, John Etchemendy, Solomon Feferman, Peter Godfrey-Smith, Yair Guttmann, Richard Rorty, Patrick Suppes, and Kenneth Taylor. In addition, I am also indebted to Henry Allison, Frederick Beiser, Domenico Bertoloni Meli, Paul Boghossian, William Boos, Tyler Burge, John Carriero, André Carus, Donald Davidson, Michael Dickson, Matthew Frank, Hartry Field, Arthur Fine, Alan Gabbey, Anil Gupta, Michael Heidelberger, David Kaplan, Noretta Koertge, Mark Lange, Elisabeth Lloyd, Edwin McCann, Thomas Nagel, William Newman, Calvin Normore, David Papineau, Paolo Parrini, Paul Pojman, Alan Richardson, Simon Saunders, Stephen Schiffer, Richard Sorrenson, Barry Stroud, Daniel Sutherland, William Talbott, Scott Tanona, Carol Voeller, Michael Williams, Mark Wilson, Rasmus Winther, John Worrall, and, as always, Graciela De Pierris. Finally, I would like to acknowledge James Mattingly for his work in preparing the index.

Note on references. In the bibliography I frequently list both an original edition and a second edition, reprinting, or translation. In citations I then give dates for both the original edition and the later one. In such cases I often give page numbers for both editions: thus "Carnap (1934/37, pp. 245–6/pp. 317–9)," for example, gives page numbers for both the original (1934) German edition and the (1937) English translation of *Logical Syntax of Language*. In some cases I give only one set of page numbers, which then refer to the later edition: thus "Quine (1951/53, pp. 42–3)," for example, gives page numbers for the (1953) reprinting of "Two Dogmas of Empiricism" in *From a Logical Point of View*.

PART ONE

The Lectures

I

The Idea of a Scientific Philosophy

My theme is the relationship between science and philosophy. The two have been intimately related, of course, throughout our intellectual history. They were born together in the Greece of the sixth through third centuries before Christ, and flowered together once again in the late medieval, renaissance, and early modern periods of the thirteenth through seventeenth centuries, which ushered in the rise of both modern science and modern philosophy as we practice them today. Unlike today, however, in both of these earlier periods there was as yet no sharp differentiation between philosophy and the sciences. Just as the schools of Plato and Aristotle made fundamental contributions to mathematics, astronomy, biology, natural history, and meteorology, as well as to philosophy, so such early modern thinkers as Descartes and Leibniz, for example, made similarly fundamental contributions both to what we now call philosophy and to the emerging new sciences of the time. That what we now call physics was still called natural philosophy at this time is a very clear indication of the absence of a sharp distinction between them.

Soon thereafter, however, the boundary lines familiar to us today began to form. To be sure, there continued to be individuals – such as Hermann von Helmholtz, Ernst Mach, and Henri Poincaré in the nineteenth century, for example – who made important contributions to both areas. And even in the vastly more specialized climate of the twentieth century, scientists whose work has had a particularly revolutionary character have continued to be involved with fundamental philosophical problems as well. In the case of Albert Einstein, for example, there is a volume of the Library of Living Philosophers devoted to him (alongside of such figures as John Dewey, George Santayana, Bertrand Russell, Ernst Cassirer, Karl Jaspers, Rudolf Carnap, Martin Buber, C. I. Lewis, Karl Popper, Gabriel Marcel, and W. V. Quine), entitled *Albert Einstein: Philosopher-Scientist*.[1] Nevertheless, the professional

---

1. Schilpp (1949).

boundaries were now clear – Helmholtz, Mach, and Poincaré were clearly professional scientists rather than professional philosophers, for example – as well as the intellectual boundaries: in the Library of Living Philosophers Einstein's is the only volume entitled *Philosopher-Scientist* (a label that would never have appeared, as such, in the time of Descartes and Leibniz, say).

With the formation of these now familiar boundary lines came new intellectual problems, particularly for those who were now professional philosophers. Since philosophy was now clearly demarcated from science, at least professionally, what should be its relation to the sciences? Should it continue to maintain very close relations to the natural and mathematical sciences, as it did in the time of Descartes and Leibniz, say, or should it rather forsake these ties in favor of closer connections to more humanistic disciplines such as history, politics, religion, or the arts? (For one important aspect of the increasing specialization of the nineteenth and twentieth centuries is that it now appeared difficult, if not impossible, to maintain equally close ties with both.) Should philosophy, despite its professional demarcation from the sciences, nevertheless strive to imitate them intellectually? Should it strive, for example, to assimilate itself within the sciences, as a branch of psychology, say, or mathematical logic? Or, failing this, should it at least strive to make itself "scientific," by replacing the traditional endless strife of metaphysical systems with a new approach to philosophical problems capable of achieving the same degree of progress, and the same degree of consensus, that are found in the sciences themselves?

By considering some of the key historical episodes in the development of this situation, these are the questions I hope to shed light on here. In particular, I will consider a number of answers that have been proposed to these questions within a tradition that became known as "scientific philosophy." Although my focus is on the special relationship between philosophy and the sciences, our discussion will, I hope, have more general implications as well. For the question of such a relationship has become a troubling one within the humanities more generally during the same period, and for some of the same reasons. It was also in the context of the increasing specialization and professionalization of the late nineteenth century, for example, that the rift C. P. Snow later characterized as an opposition between "the two cultures" first arose, in the differentiation, within *Wissenschaft* in general (which term has a much more general meaning in German than our own term "science"), between the *Naturwissenschaften* and the *Geisteswissenschaften*. And the sense of tension and unease in the relationship between these two areas of course persists today. For many in the humanities, for example, Thomas Kuhn's *The*

Structure of Scientific Revolutions brought a welcome sense of relief and vindication. Since Kuhn has shown that science has no special or privileged intellectual standing after all, so the argument goes, but is simply one more "disciplinary matrix" or intellectual community within our culture, we in the humanities need no longer worry about the presence or absence of a "scientific" foundation for our own disciplines. Indeed, since it is we in the humanities who have "culture" for our particular object of study, it is we, and not the scientists themselves, who are most competent to discuss the question of the ultimate "justification" or "legitimation" of the sciences. This line of thought has of course led inevitably to the recent "science wars," where even some scientists now feel themselves both intellectually and professionally on the defensive.

The concept of a "scientific philosophy" (or *wissenschaftliche Philosophie*) first developed in the mid nineteenth century, as a reaction against what was viewed as the excessively speculative and metaphysical character of post-Kantian German idealism. One of the primary intellectual models of this movement was a celebrated address by Hermann von Helmholtz, "Über das Sehen des Menschen," delivered at the dedication of a monument to Kant at Königsberg in 1855 (Helmholtz at the time was professor of physiology at Königsberg.) Helmholtz begins by asking, on behalf of the audience, why a natural scientist like himself is speaking in honor of a philosopher. This question only arises, he says, because of the current deplorable climate of enmity and mutual distrust between the two fields – a climate which is due, in Helmhotz's opinion, to the entirely speculative system of *Naturphilosophie* that Schelling and Hegel have erected wholly independent of, and even in open hostility towards, the actual positive results of the natural sciences. What Helmholtz is now recommending, however, is a return to the close cooperation between the two fields exemplified in the work of Kant, who himself made significant contributions to natural science (in his nebular hypothesis put forward in 1755), and, in general, "stood in relation to the natural sciences together with the natural scientists on precisely the same fundamental principles."[2] And it was this recommendation that was enthusiastically embraced within the emerging "back to Kant!" movement, where it led to the idea that all metaphysics should be replaced by the new discipline of "epistemology" or

---

2. See Helmholtz (1865/1903, vol. I, p. 88).

"theory of knowledge" (*Erkenntnistheorie*), so that philosophy itself would now become "scientific." This movement then found its culmination in a new journal, the *Vierteljahrsschrift für wissenschaftliche Philosophie*, founded in 1877.[3]

Now the charge that Helmholtz – and, following him, the rest of this "back to Kant!" movement for a scientific philosophy – leveled against the *Natur-philosphie* of the early nineteenth century is no doubt fundamentally unfair. Not only were the *Naturphilosophen* trying to respond intellectually to some of the key scientific developments of their time, developments in chemistry, electricity and magnetism, and energetics, for example, but it is also arguable that some of these key developments, including Helmholtz's own formulation of the conservation of energy in 1847, were themselves significantly influenced by *Naturphilosophie*.[4] But what is of primary interest, from our present point of view, is the nature and character of the new scientific philosophy that was now being explicitly opposed to *Naturphilosophie* in particular and post-Kantian idealism in general. What relation is philosophy now supposed to bear to the sciences, and what does it mean for philosophy to become scientific in this way? What exactly is being recommended when we are told that philosophy should stand "in relation to the natural sciences together with the natural scientists on precisely the same fundamental principles"?

For Helmholtz himself this means that philosophy – that is, epistemology or the theory of knowledge – should work in cooperation with the latest psycho-physiological research in inquiring into the nature of the representations of our senses, and the relationship between these representations and the actual world to which they correspond. And it is for this reason that the body of his 1855 address is occupied almost exclusively with reporting on some of his own work in the psycho-physiology of vision, which he had begun as a student under Johannes Müller in Berlin. As he makes clear in his most mature presentation of his epistemology in "The Facts in Perception" of 1878, it is Helmholtz's view that philosophy considers the relationship between our representations and the external world from the inside out, as it were, while natural science – in this case psycho-physiology – considers the very same relationship from the outside in. Philosophy thus consider our knowledge

---

3. For discussion of the development of neo-Kantianism and the "back to Kant!" movement see Köhnke (1986/91).

4. For discussion of the discovery of the conservation of energy, in particular, with some remarks about the influence of *Naturphilosophie* on a variety of scientific developments of the period, see Kuhn (1959/77).

from the mental or psychological side, while natural science considers it from the physical or physiological side:

> The fundamental problem, which that time placed at the beginning of all science, was that of the theory of knowledge: 'What is truth in our intuition and thought? In what sense do our representations correspond to actuality?' Philosophy and natural science encounter this problem from two opposite sides; it is a common task of both. The first, which considers the mental side, seeks to separate out from our knowledge and representation what originates from the influences of the physical world, in order purely to establish what belongs to the mind's own activity. Natural science, by contrast, seeks to separate off what is definition, designation, form of representation, and hypothesis, in order purely to retain what belongs to the world of actuality, whose laws it seeks.[5]

In both cases, however, our inquiry rests wholly and completely on the latest empirical findings of psychological and physiological research, and so, in the end, philosophy, for Helmholtz, is itself an empirical natural science – a branch of empirical psychology. In this way, Helmholtz anticipates the conception, popular in some circles today, that philosophy should become absorbed into cognitive psychology.

Helmholtz is thus being somewhat disingenuous in invoking the authority of Kant, and, more particularly, in recommending that we return to the close relationship between philosophy and natural science as Kant envisioned it. For Kant himself held that philosophy has a special "transcendental" status that sharply differentiates it from all empirical science, including, and indeed especially, empirical psychology. For example, in leading up to a characterization of "transcendental logic" – another name for what he is here calling "transcendental philosophy" – Kant explains that logic "as pure ... has no empirical principles, and hence borrows nothing (as one has sometimes supposed) from psychology, which thus as no influence at all on the canon of the understanding." And a few pages later Kant makes a remark which, as he says, "extends its influence over all the following considerations, and which one must bear well in mind":

---

5. See Hertz and Schlick (1921/77, p. 111/pp. 117–8).

> [N]ot every a priori cognition should be called transcendental,
> but only that through which we know that and how certain rep-
> resentations (intuitions or concepts) are applied wholly a
> priori, or are possible (that is, [through which we know] the
> possibility or the a priori employment of the cognition). There-
> fore, neither space nor any a priori geometrical determination
> thereof is a transcendental representation, but what can alone be
> called transcendental is the knowledge that these representa-
> tions are not at all of empirical origin, and the possibility that
> they can nevertheless relate a priori to objects of experience.
> (A56/B80-1)[6]

Philosophy, as a "transcendental" inquiry, is not only distinct from all empirical science, but it is also distinct from those elements of pure a priori
knowledge, such as geometry, for example, which are present in the sciences
themselves. Whereas each of the first-level sciences, whether empirical or a
priori, has its own characteristic objects, philosophy, as a second-level or
meta-level discipline, has no such objects of its own, but rather concerns the
nature and possibility of our *representations* of these objects. The distinctive
subject matter of philosophy is thus our *knowledge* of these first-level objects.
As Kant puts it elsewhere (B25): "I term all cognition transcendental which
occupies itself in general, not so much with objects, but rather with our mode
of cognition of objects, in so far as this is supposed to be possible a priori."

   Indeed, this Kantian distinction between first-level scientific inquiries and
the distinctively philosophical "transcendental" inquiry is actually the historical source for the intellectual differentiation between philosophy and the
sciences which is now familiar today. The rationalist philosophers of the seventeenth century, such as Descartes and Leibniz, had distinguished between
physics or natural philosophy, on the one hand, and metaphysics or "first philosophy," on the other. But they by no means meant by this a distinction
between two essentially different types or levels of inquiry in the Kantian
sense. Rather, just as physics or natural philosophy studies the visible or corporeal part of the universe, metaphysics or "first philosophy" studies the
invisible or incorporeal part – that is, God and the soul. And it is precisely by
articulating the structure of the invisible or incorporeal part of the universe
that "first philosophy" or metaphysics can then provide a rational foundation

---

6. All references to the *Critique of Pure Reason* are given by the standard pagination of the first (A)
and second (B) editions. The earlier cited remark about the purity of logic occurs at A54/B78.

for physics: a rational foundation, that is, for the new mathematical natural philosophy to which these rationalist philosophers were themselves making vitally important contributions. In their intellectual enterprise – that of articulating the rational structure of the universe as a whole – "first philosophy" and natural philosophy are thus entirely continuous.

Kant, by contrast, is breaking decisively with this tradition. For him, (theoretical) rational knowledge of incorporeal objects, and, in particular, of God and the soul, is completely impossible for us.[7] The only possible objects of our human knowledge are "appearances," that is, objects in space and time interacting with one another in accordance with the causal laws of the new mathematical natural science. So metaphysics or "first philosophy" in the sense of seventeenth century rationalism is also entirely impossible. What can replace this hopeless enterprise, however, is a new "transcendental" inquiry into the *conditions of possibility* of our first-level knowledge of objects in space and time (the only genuine objects of knowledge there now are) supplied by mathematical natural science. In this way, by renouncing all claims on the supersensible, and redirecting our attention rather to the necessary conditions which make possible natural scientific knowledge (the only genuine knowledge of objects we now have), philosophy or metaphysics can finally leave behind the "mock combats" of the schools, and itself enter into "the secure path of a science."[8] In this way, too, philosophy can give a rational foundation, but in an entirely new sense, for the natural scientific knowledge whose conditions of possibility it investigates.

Kant's new concern with what he calls conditions of possibility is fundamentally shaped by the scientific context of the eighteenth century – the age of the triumph of Newtonianism. The rationalist thinkers of the seventeenth century had acted as contributors, apologists, and propagandists for the mechanical natural philosophy: the inspiring vision, fueled by Copernicanism and the example of Galileo, of a precise mathematical description of all of the phenomena of nature under a single set of mathematical laws uniting the earth and the heavens, to be achieved by an atomistic or corpuscular theory of matter that reduced all natural changes to the motions and mutual impacts of the constituent particles. But the rationalist defense of the mechanical philosophy was also highly programmatic in character, in so far as nothing even approaching such a unified mathematical description was actually produced.

---

7. Kant does believe, however, that we have *practical* access to the supersensible through our experience of the moral law. This is the point of his famous dictum that he "had to destroy *knowledge* [*Wissen*] in order to procure a place for *belief* [*Glauben*]" (Bxxx).
8. See note 11 below.

Only with Newton, in fact, was even the very first step of this program, the synthesis, under a common set of dynamical laws, of celestial astronomy with aspects of terrestrial physics (namely, those due to gravity) actually achieved.

Yet this Newtonian synthesis also raised serious conceptual problems. In the first place, it portrayed gravitational interaction as an immediate action at a distance across arbitrarily large empty spaces, thereby breaking with the fundamental tenet of the mechanical philosophy that all action should take place by contact. So it appeared that we were either left with a commitment to just the kind of "occult quality" that the mechanical philosophy had dedicated itself to overcome (a primitive *attraction*), or forced to acquiesce in a merely empirical and phenomenological physics that renounced all inquiry into "true causes." In the second place, however, and even more seriously, Newton's physics was deliberately and self-consciously erected on the concepts of absolute space, time, and motion. These concepts were also entirely unacceptable to the seventeenth century rationalism which fueled the mechanical philosophy – on the basis of the sharp division, noted above, between the visible or corporeal part of the universe and the invisible or incorporeal part. For absolute space and time appeared to occupy a completely untenable intermediate position, as incorporeal but nonetheless physical. (And it was precisely this intermediate status that led to the traditional puzzles about the relationships between space, time, and divinity.)

In the eighteenth century, then, Newton's physics was an unqualified success in both mathematical and empirical terms, but there remained serious conceptual problems concerning whether and how this brilliantly successful theory actually made rational sense. Kant's problem, accordingly, was not to sketch a program for a new mathematical physics, but rather to explain how our actual mathematical physics, the mathematical physics of Newton, was itself possible in the first place. And his answer, in the briefest possible terms, is that the concepts of space, time, motion, action, and force do not function to describe a metaphysical realm of entities or "true causes" lying behind the phenomena. Nor are they simply abstractions from our experience, which we can then apply to the phenomena because we have already found them there. Rather, such concepts as space, time, motion, action, and force are a priori forms or constructions of our own, on the basis of which alone we can coherently order the phenomena of nature into a unified and law governed spatiotemporal totality. Absolute space, for example, is not a metaphysical entity, a great empty "container," existing behind the phenomena independently of all material content. Yet, as the success of Newton's physics dramatically shows, we cannot simply dispense with it either, as an empty concept with no empirical application. On the contrary, we apply this concept empirically in constructing *approximations* to a privileged frame of reference at absolute rest – as a first approximation, for example, Newton, in the *Principia*, has himself constructed the center of mass frame of the solar system.[9] And action at a distance, despite its violation of the strictures of the mechanical philosophy, is a similarly necessary concept, in its realization by universal gravitation, for empirically establishing the temporal *simultaneity* of arbitrarily distant events.[10]

For Kant, then, Newtonian physics is not simply a pragmatically successful scheme for ordering and predicting the phenomena. It also counts as a model or paradigm for the coherent rational comprehension of nature, because it injects a priori forms, constructions, or categories of our own – which, for Kant, express universal capacities of the human mind – into our experience of nature. It is in this way, and this way alone, for Kant, that we can rationally explain how such pragmatic empirical success is actually possible in the first place. And it is in this way, too, that we can finally set what was previously called metaphysics or "first philosophy" onto "the secure path of a science," where we leave behind its former condition of "random groping," "mock combat," and utter lack of unanimity:

> With respect to the question of unanimity among the adherents of metaphysics in their assertions, it is still so far from this that it is rather a battle ground, which seems to be quite peculiarly destined to exert its forces in mock combats, and in which no combatant has ever yet been able to win even the smallest amount of ground, and to base on his victory an enduring possession. There is therefore no doubt that its procedure has, until

---

9. For Kant this procedure is then continued indefinitely: from the center of mass of the solar system to the center of mass of the Milky Way galaxy, from there to the center of mass of a system of such galaxies, from there to the center of mass of a system of such systems, and so on. For discussion of Kant's analysis of Newtonian physics with special attention to the problem of absolute space, time, and motion see Friedman (1992, chapters 3 and 4).

10. Kant's Third Analogy of Experience depicts the most general conditions for establishing the temporal relation of simultaneity (whereas the First and Second deal with duration and succession respectively). In the *Metaphysical Foundations of Natural Science* (1786) the Third Analogy is realized or instantiated by the Newtonian third law of motion – the equality of action and reaction. Universal gravitation is then an even more specific realization of this last principle. For further discussion see Friedman (1992), and compare also Part Two, section 2 below (in particular note 23).

now, been a merely random groping, and, what is worst of all, among mere concepts. (Bxiv–v)[11]

By renouncing all claims to a special, as it were, "supernatural" domain of objects of its own, and rather confining itself to the articulation of the necessary conditions of possibility of the natural and mathematical sciences, philosophy, although not itself a science in the same sense, can nevertheless achieve stable and definitive results, and thus finally become, in this sense, scientific.

Let us now move forward to the year 1921, the centenary year of Hermann von Helmholtz's birth. Helmholtz's remarkable and wide-ranging scientific achievements, in energetics, physiological psychology, the foundations of geometry, electrodynamics, and epistemology, were celebrated in a variety of memorial addresses, journal issues, monographs, and the like, including an address given by the philosopher Moritz Schlick at the University of Berlin, entitled "Helmholtz als Erkenntnistheoritiker." Schlick had earlier earned a doctorate in theoretical physics under Max Planck at Berlin, but soon thereafter decided to pursue a career in philosophy instead. His Habilitation in philosophy, on "The Essence of Truth in Modern Logic," was published in 1910 in the *Vierteljahrsschrift für wissenschaftliche Philosophie und Soziologie* (which the *Vierteljahrsschrift für wissenschaftliche Philosophie* had turned into in 1901). Schlick then became the leading philosophical proponent and expositor of Einstein's new theory of relativity with the publication of his extremely influential monograph, *Space and Time in Contemporary Physics*, which went through four editions between 1917 and 1922. In 1922, largely on the strength of his work on the philosophical significance of the theory of relativity, which had been enthusiastically endorsed by Einstein himself, Schlick was named to the Chair for the Philosophy of the Inductive Sciences previously occupied by the scientists Ernst Mach and Ludwig Boltzmann at the University of Vienna, where he became the leader and guiding spirit of what we now know as the Vienna Circle of logical positivists. We might say, in this sense, that Schlick was the very first professional scientific philosopher.

---

11. In the preceding pages Kant describes how logic, mathematics, and natural science have been placed on "the secure path of a science," and in the succeeding pages he explains how metaphysics can now be placed on this path in a similar fashion ("by a single and sudden revolution") through precisely his own revolution in philosophy.

Just as Helmholtz, in his dedicatory address on human vision of 1855, invokes the authority of Kant on behalf of his own conception of scientific philosophy, Schlick, in his memorial address of 1921, invokes the authority of Helmholtz, and the example of Helmholtz's 1855 address, on behalf of his conception of the relation of science to philosophy. All great scientists, according to Schlick, "think every problem with which they are concerned up to the end, and the end of every problem lies in philosophy." By beginning with special problems of the special sciences, we ascend step by step to "the ultimate attainable principles ... which, because of their generality, no longer belong to any special science, but rather lie beyond them in the sphere of the general theory of science, in philosophy, in the theory of knowledge." Physics, for example, "penetrates only ... to certain ultimate concepts and presuppositions – such as space, time, causality ... whose illumination and justification must remain left to philosophy."[12] Yet, as Schlick makes clear elsewhere, this does not mean that philosophy is a separate discipline from the sciences:

> Philosophy is not an independent science that would be placed next to or above the individual disciplines. Rather, what is philosophical is found *in* all the sciences as their true soul, in virtue of which they first become sciences at all. Every particular field of knowledge, every special form of knowing, presupposes the most general principles into which it flows and without which it would not be knowledge. Philosophy is nothing other than the system of these principles, which branches out and penetrates the system of all knowledge and thereby gives it stability; it therefore has its home in all the sciences.[13]

Philosophy, for Schlick, does not have any special relation to psychology. It is not, as it was for Helmholtz, especially concerned with the psycho-physiological mechanisms of human sense perception. Philosophy is rather concerned with the foundations or ultimate principles of each and every science, whereby each of the special sciences takes its own particular place in the total system of knowledge. Philosophy, we might say, supplies the foundational and systematic core of each of the special sciences; it is neither a meta-science (as it was for Kant) nor particularly connected with any individual special science (as it was for Helmholtz).

---

12. Schlick (1922/78, pp. 29–30/p. 335).
13. Preface to *General Theory of Knowledge*: Schlick (1918/85, p. vii/p. v).

Nevertheless, the ultimate scientific principles with which Schlick was himself especially concerned, as he indicates in his 1921 address, are the principles of Einstein's new physics – principles of "space, time, causality." And, with respect to these principles, it is not too much to say that Schlick aimed to do for Einstein's physics what Kant had done for Newton's, namely, to explain and exhibit the special features of this physics that make it a model or paradigm of coherent rational knowledge of nature. One central implication of this new physics, however, is that Kant's conception of natural knowledge, as framed by universal forms or categories of the human mind, taken to be rigidly fixed for all time, cannot, after all, be correct. For it is precisely the Newtonian conceptions of space, time, motion, and interaction that Einstein has now rejected and replaced; and so these particular ultimate principles in no way have the status Kant had attributed to them. Yet this does not mean, as Ernst Mach would have it, for example, that physics can simply be abstracted or generated from sense experience and from sense experience alone. On the contrary, we still need superordinate and highly mathematical first principles in physics – principles that must be injected into our experience of nature before such experience can teach us anything at all. But these principles do not express a priori fixed features of the human mind, as Kant would have it. They rather have the status of what Henri Poincaré, in his own work on the philosophical foundations of geometry, called "conventions" – free *choices* of our own needed to bridge the irreducible gulf between our crude and approximate sensory experience and our precise mathematical descriptions of nature.

So things stood for Schlick in 1922, when he moved to the University of Vienna. In the discussions of what we now know as the Vienna Circle, however, Schlick's earlier conception of philosophy, as the foundational core, as it were, of all the special sciences, underwent a radical transformation. For one of the first orders of business of the Circle was to assimilate and appropriate the new advances in mathematical logic due to Gottlob Frege and Bertrand Russell, as these advances were philosophically articulated and interpreted by the *Tractatus Logico-Philosophicus* (1922) of the young Ludwig Wittgenstein. But here Schlick and the Circle encountered the following ideas:

> The totality of true propositions is the whole of natural science (or the totality of the natural sciences).

> Philosophy is not one of the natural sciences.
> (The word "philosophy" must mean something that stands above or below the natural sciences, not beside them.)

> The aim of philosophy is the logical clarification of thoughts.
> Philosophy is not a doctrine but an activity.
> A philosophical work consists essentially of elucidations.
> Philosophy does not result in "philosophical propositions,"
> but rather in propositions becoming clear.[14]

A proper understanding of the new mathematical logic shows, according to the *Tractatus*, that the only meaningful propositions – propositions that can be meaningfully said to be true or false – are those of the individual natural sciences. And there can be no "ultimate principles" of these sciences whose articulation and formulation would belong, as its special province, to philosophy. All philosophy can do is analyze the logical form or logical structure of the propositions of the special sciences, whereby it issues in no propositions – no "ultimate principles" – of its own, but simply in the activity of logical clarification itself, an activity that does not involve the formulation or articulation of further meaningful propositions in turn.

Indeed, it is precisely the misconstrual of philosophy as a body of doctrine that is responsible for the confusions, and in fact utter nonsense, of traditional metaphysics:

> The correct method of philosophy would properly be the following: To say nothing but what can be said, that is, the propositions of natural science – and thus something that has nothing to do with philosophy – and then, whenever another wanted to say something metaphysical, to demonstrate to him that he had given no meaning to certain signs in his propositions.[15]

If the scientific philosophers of the Vienna Circle truly wanted to avoid metaphysics, it now appeared, they would also have to give up the idea that philosophy could be a science in *any* sense. Philosophy, on Wittgenstein's conception, is not even a theoretical discipline at all, but simply the (nontheoretical) *activity* of logical analysis.

Here the Vienna Circle, and their ideal of a scientific philosophy, were clearly caught in a most uncomfortable position. To the rescue, as it were, came Rudolf Carnap, ten years younger than Schlick, who had joined the Circle at the University of Vienna in 1926. Carnap, like Schlick, had focussed in

---

14. Wittgenstein (1922, §§ 4.111–4.1112)
15. Ibid., § 6.53.

his graduate education on theoretical physics which he studied under Max Wien at the University of Jena, where he studied Kantian and neo-Kantian philosophy as well, and, perhaps most significantly, also the new mathematical logic under Gottlob Frege. Since the physics faculty found his plans for a doctoral dissertation on the axiomatic foundations of relativity theory to be too philosophical, Carnap ultimately wrote an interdisciplinary dissertation combining physics, mathematics, logic, and philosophy, which was published in 1922. In this dissertation he arrived at an analysis of the new concepts of space and geometry that were largely in harmony with, and indeed significantly influenced by, the conclusions Schlick had reached in 1917. He agreed with Schlick, in particular, that Kant's original conception of the fixed and necessary status of Euclidean geometry must be replaced by Poincaré's idea that the geometry of physical space rests rather on a *convention* freely chosen on the basis of the simplicity of our overall system of geometry plus physics.[16]

But Carnap's original contributions to the Vienna Circle concerned mathematical logic. For it was he, at least among the professionally philosophical members, who had the deepest understanding of, and appreciation for, the rapidly accumulating results of this new discipline. In the late 1920s, in particular, he became deeply immersed in the program of David Hilbert, perhaps the greatest mathematician of the twentieth century, to create a new logical discipline called *metamathematics*. Here we view logic and mathematics purely formally, as mere syntactical systems of sentences and proofs, and we then apply this new point of view in investigating mathematically the logical – or rather meta-logical – relations within such a system: we investigate notions like derivability, definability, consistency, completeness, and so on. In his *Logical Syntax of Language* of 1934 Carnap urged that we should extend Hilbert's method from logic to the whole of philosophy. Scientific philosophy should now become *Wissenschaftslogik* – the meta-logical investigation of the logical structures and relations of the total language of science. And in this way, as Carnap explicitly argues, we finally have an alternative to what he took to be the "mystical," and therefore fundamentally unscientific conception of philosophy of the *Tractatus*, according to which logical form or syntax is ineffable and unarticulable – describable by no meaningful propositions of its own. On the contrary, *Wissenschaftslogik*, the meta-logical investigation of the logical syntax of scientific language, is itself a perfectly precise and rigorous system of logico-mathematical propositions. Philosophy is a branch of mathematical logic:

---

16. Carnap (1922). For discussion see Friedman (1999, chapter 2).

The alleged peculiarly philosophical point of view, from which the objects of science are supposed to be considered, is abolished, just as the alleged peculiarly philosophical stratum of objects was already previously eliminated. Aside from the questions of the individual special sciences, the only questions that remain as genuinely scientific questions are those of the logical analysis of science – its sentences, concepts, theories, etc. We will call this complex of questions *Wissenschaftslogik*. ... Taking the place of the inextricable tangle of problems that is known as philosophy is *Wissenschaftslogik*. Whether, on the basis of this conception, the designation "philosophy" or "scientific philosophy" should be applied to this remainder, is a question of expedience, which is not to be decided here.[17]

For Carnap, then, philosophy or logical analysis is a meta-science, as it was for Kant. In contrast to Kant, however, it is also a branch or part of science as well – this time a branch of *formal* or a priori (as opposed to empirical) science.

We obtain a fundamentally new understanding of the character of philosophical problems in this way. Traditional philosophical debates, such as the debate between "realist" and "idealist" conceptions of the external world, for example, do not concern matters of fact concerning which one can possibly be either correct or incorrect. Viewed in this way, as the history of metaphysics amply demonstrates, there is absolutely no possibility of resolution. Such philosophical "doctrines" should rather be viewed as *proposals* – as proposals to construct the total language of science in one way rather than another. The "idealist" proposes to formulate the language of science beginning with a basis in sense-data or private experience, for example, while the "realist" proposes to begin with a basis in the concepts of physics. Both languages are perfectly possible, for both can be represented as alternative formal systems or axiomatizations, as it were, within logical syntax. And each such language or linguistic framework, axiomatized in its own particular way, yields its own particular standards of logical correctness and thus truth – standards that define what Carnap calls *internal* questions relative to a given language or linguistic framework. The question of which language or linguistic framework we should adopt, by contrast, is an *external* question. And here no question of

---

17. Carnap (1934/37, section 72). Carnap sharply differentiates himself from Wittgenstein's doctrine of the inexpressibility of logical syntax – explicitly taking issue with the ideas cited in note 14 above – in the following section.

correctness or truth can arise at all, but only purely conventional or pragmatic questions of suitability or appropriateness for one or another given
purpose.[18]

Let us now move forward once again, this time to the year 1962. Many of the
leading members of the logical positivist movement, including Carnap, have
long since become established in the United States, where they emigrated to
escape the Nazi regime in the mid to late 1930s. (Schlick, although he had vis-
ited the United States in the late twenties and early thirties, was murdered at
the University of Vienna by a former student in 1936.) In the comfortable climate of American pragmatism and common-sense empiricism, however, the
positivists lost much of their revolutionary fervor. No longer militantly cru-
sading for a reform of philosophy as a whole, for a new type of scientific
philosophy, they instead became respectable (and domesticated) practitioners of a new sub-discipline called philosophy of science. And, despite the
impressive gains in clarity of some of the logical analyses thereby produced,
this sub-discipline had reached a relatively unexciting period of stasis by the
late 1950s and early 1960s. In 1962, however, Carnap and the American pragmatist philosopher Charles Morris (who had been instrumental in bringing
many of the positivists to the United States) published, under their joint edi-
torship, a new volume of the *International Encyclopedia of Unified Science*
(which had become the official monograph series of the logical positivist
movement in 1938) by the young American physicist turned historian of science, Thomas Kuhn. This, of course, was *The Structure of Scientific
Revolutions*. We know, from Carnap's correspondence with Kuhn at the time,
as well as from his own unpublished notes, that Carnap himself was
extremely enthusiastic about Kuhn's work.[19]

There is considerable irony in this, of course, for *The Structure of Scientific*

---

18. The distinction between internal and external questions is first made explicit in Carnap (1950/
56), but the fundamental idea is already clearly present in *Logical Syntax*. There it is applied primarily to the current dispute in the foundations of mathematics between logicism, formalism,
and intuitionism – each of which is reinterpreted as a proposal to formulate the total language of
science in accordance with one or another system of logical rules (with or without the law of
excluded middle, for example). The "realist"/"idealist" debate is also reinterpreted in this way
during the same period, which actually extends back to the *Aufbau* (1928). For further discussion
see Friedman (1999, chapter 9).
19. For discussion (together with reproductions) of these materials see Reisch (1991), Earman
(1993).

Revolutions is often taken to represent the death-knell of the logical positivist philosophy of science represented by Carnap. Indeed Kuhn himself, in a state of blissful but perhaps forgivable innocence of the positivists's early work on the revolutionary import of the theory of relativity, uses that very theory to make his own case, on behalf of his conception of "the nature and necessity of scientific revolutions," *against* what he calls "early logical positivism."[20] But I do not want to dwell further here on this particular irony of history. Instead, I want to explore what we can now learn about the idea of a scientific philosophy, against the philosophical background we have briefly sketched above, from Kuhn's theory of scientific revolutions. In particular, I want to use Kuhn's theory of scientific revolutions, supplemented by a parallel consideration of the concurrent developments taking place within the discipline of philosophy, to show that none of the conceptions of the relationship between philosophy and the sciences we have so far considered is fully satisfactory – although each, as we shall see, contains a part of the truth, and, taken together in this way, they can serve to point us in a more fruitful direction.

The first point to notice, in this connection, is that, although Kuhn's book, as I noted at the beginning, is often taken to support the idea that there is no fundamental difference between the sciences and other areas of intellectual and cultural life, the book actually begins by delineating just such a fundamental difference. In the first full chapter, entitled "The Route to Normal Science," Kuhn outlines how the disciplines he calls sciences – or better, mature sciences – emerge from the "pre-paradigm" state. Such a transformation occurs, according to Kuhn, when a number of diverse and competing schools of thought within a discipline or area of inquiry are replaced by a single model or paradigm that is "universally received" within this area of inquiry as the basis for a "firm research consensus" – a consensus or agreement on a single set of rules of the game, as it were, which set the parameters of inquiry for all practitioners of the discipline from that point on (at least for a time).[21] It is only when such an at least relatively enduring consensus is achieved that we have what Kuhn calls normal science, and it is only against the background of such an already existing state of normal science that we can then have a scientific revolution – which occurs precisely when one such enduring stable consensus is replaced by a different one. Mathematics and

---

20. This famous discussion of the relationship between special relativistic and Newtonian physics occurs in Kuhn (1962/70, chapter 9), where Kuhn rejects the view, "closely associated with early logical positivism," that the latter theory can be logically derived from the former. (Kuhn does not actually cite any of the early logical positivists by name here.)
21. Here, and in the rest of this paragraph, see Kuhn (1962/70, chapter 2).

astronomy, Kuhn suggests, reached this state of normal science in antiquity; what we now call mathematical physics, however, reached it only with the great events of the sixteenth and seventeenth centuries that culminated in the work of Newton; chemistry achieved this status even later; biology still later; and so on. It remains an open question, Kuhn adds, "what parts of social science have yet acquired such paradigms at all." And in the humanities, he scarcely needs to add, it is completely clear that a state of normal science is never achieved: perpetual competition between mutually opposing schools of thought is the name of the game.

I want now to consider the particular case, among the humanities, of philosophy. I want to consider, more specifically, how the historical evolution of philosophy as a discipline, although very different from that of the sciences, is nonetheless intimately connected with the development of the sciences. What I find unfortunately lacking in Kuhn is precisely such a parallel historical treatment of philosophy. Indeed, in Kuhn's book philosophy is treated quite *ahistorically,* and in an entirely partisan and polemical manner, as what everyone else supposedly thought before Kuhn himself arrived on the scene.[22]

In philosophy the most we ever achieve is temporary consensus on which figures or doctrines set the philosophical agenda – for the moment, that is, and within a relatively circumscribed setting. With the publication of the *Critique of Pure Reason* in 1781, for example, Kant set the agenda for German philosophy. This did not result, however, in anything like a paradigm in Kuhn's sense, a single set of generally agreed upon rules of inquiry.[23] On the contrary, we immediately saw the rise of differing, and mutually hostile interpretations of Kantian transcendental philosophy, culminating in new versions of such philosophy, the systems of post-Kantian German idealism, that radically revised and even rejected some of Kant's most basic principles. This movement, as we have seen, was then countered, in turn, by a reaction in favor of a new type of "scientific philosophy" aiming to return to what it took to be most important in Kant. And scientific philosophy itself fragmented

---

22. A particularly striking and *exuberant example of this occurs* in Kuhn (1962/70, chapter 10, p. 126): "But is sensory experience fixed and neutral? Are theories simply man-made interpretations of given data? The epistemological viewpoint that has most often guided Western philosophy for three centuries dictates an immediate and unequivocal, Yes!"

23. Here we have an interesting application of the well-known ambiguity of the term "paradigm," where it can mean both an exemplary work or contribution and a set of common standards for "normal" research consequent on the acceptance of a given work or contribution as exemplary. In contrast to the (mature) sciences in Kuhn's sense, in philosophy (and in the humanities more generally) the first sense of "paradigm" is typically *not* accompanied by the second. (I am indebted to discussion with Michael Dickson on this point.)

once again, into neo-Kantianism, on the one side, and what then became logical positivism, on the other. Never, not even within this relatively circumscribed world of post-Kantian German philosophy, is anything like a stable consensus on a common paradigm attained. We see only a constantly shifting dialectic of thought, now fragmenting and dividing, now momentarily coalescing, between similarly shifting philosophical positions and schools.

Yet the constantly shifting stream of philosophical thought is inextricably entangled, as we have also seen, with the very different evolution of scientific thought portrayed by Kuhn: an evolution which moves from a pre-paradigm state of conflicting schools to a universal research consensus on a single paradigm, proceeds in an enduring stable state of normal science for a significant time, and finally, through the accumulation of anomalies, is punctuated by a scientific revolution – a comparatively rapid transition to a second relatively stable paradigm – whereupon we return again to a state of normal science, and so on. Thus, at the same time that he set the agenda for the mechanical natural philosophy, involving a new paradigm for physics based on corpuscularianism and action by contact, Descartes simultaneously set the agenda for modern philosophy by radically revising and reorganizing the wider system of philosophical concepts and principles bequeathed to western thought by Scholasticism. And after the physical paradigm of the mechanical natural philosophy was itself radically transformed by Newton, Kant found it necessary to venture a new fundamental reorganization of knowledge, where, for the first time, the discipline of philosophy became definitively separated from the sciences, as an essentially "transcendental" inquiry aiming to explain the necessary conditions of possibility of the new paradigm for mathematical physics created by Newton. Scientific thinkers of the nineteenth century, such as Helmholtz and Poincaré, for example, finding Kant's necessary conditions of possibility too restrictive, and responding to the stresses and strains that were developing at the time within the Newtonian paradigm itself, then struggled to articulate new types of scientific philosophy. These struggles eventuated in Einstein's formulation of the theory of relativity, where a new physical paradigm for the study of space, time, motion, and interaction was finally created. This new paradigm led, in turn, to the philosophy of logical positivism.[24] And so on.

---

24. These central examples will be further discussed in the remainder of the Lectures; Part Two contains a particularly detailed discussion of the nineteenth century philosophical background to relativity.

Indeed, we can go further. For, at moments of scientific revolution, the scientific transitions themselves (the transitions to a new paradigm) are actually quite inconceivable without the parallel developments in philosophy taking place at the same time, and, as it were, on a different level. In normal science we operate within the context of a generally agreed upon framework that defines the norms and standards, the rules of the game, for a given area of inquiry. These standards, in normal science, are not themselves called into question. On the contrary, it is they, taken simply as given, which then make the problem solving activities of normal science possible in the first place. (In Carnapian terms, they constitute the rules of a given linguistic framework definitive of a given set of internal questions.) During periods of deep revolutionary change, by contrast, it is precisely such a formerly agreed upon set of standards that has now been called into question; and so we no longer have such standards at hand to call upon in motivating and supporting the transition to a new paradigm. (In Carnapian terms, we are here faced with an external question concerning the replacement of the rules of one linguistic framework by a very different set of rules.)[25] In making this kind of transition, therefore, we are no longer dealing with purely scientific questions in the same sense – that is, we are no longer operating wholly within normal science – and it is precisely here that characteristically philosophical considerations come into play. And it is precisely this, in my opinion, that we still must add to Kuhn's picture.

The adoption of the mechanical natural philosophy by the scientific thinkers of the seventeenth century, for example, was not simply motivated by the mathematical and empirical success it had achieved. For, as noted above, the reach of this new intellectual movement far exceeded its grasp. One here aimed at nothing less than a precise mathematical description of all of the phenomena of nature, to be achieved by an atomistic or corpuscular theory of matter, and nothing even approximating such an atomistic reduction was actually achieved until the late nineteenth and early twentieth centuries – when, we might add, it was achieved using entirely new and hitherto entirely unforeseen mathematical and physical concepts.[26] In the seventeenth century itself, however, this new physical paradigm remained almost entirely programmatic. And the new paradigm was motivated and sustained, especially

25. Compare note 18 above. It is because of precisely this parallel, of course, that Carnap was so enthusiastic about Kuhn's work (note 19 above). (And this, in turn, is closely connected with the actual assimilation of the theory of relativity by the "early logical positivists": see note 37 below.)
26. The necessary concepts to make atomism actually work turn out to be precisely those of relativity and quantum mechanics. Compare Part Two, section 5 below, especially note 66.

during its first fifty years, not so much by mathematical and empirical success, but by the inspiring philosophical vision of a radically new approach to the understanding of nature self-consciously crafted by Descartes and Galileo against the background of late medieval Scholasticism. Similarly, to take a second example, in the relativistic revolution wrought by Einstein purely mathematical and empirical considerations again played a decidedly secondary role. For, at the time of Einstein's initial formulation of the special theory of relativity in 1905, there was on the scene a fully developed competitor theory – the Lorentz-Fitzgerald theory of the "aether" – which was, in an important sense, both mathematically and empirically equivalent to Einstein's theory. Einstein's great innovation was rather a conceptual one – the recognition of a new item, as it were, in the space of intellectual possibilities: namely, the possibility of a relativized conception of time and simultaneity. And it was motivated and sustained by a similarly new philosophical conception developed against the background of the nineteenth century debate between Kantianism and empiricism in the philosophy of geometry – namely, the insight of Henri Poincaré (whom Einstein was intensively reading at the time, and who was also deeply concerned with the foundations of Lorentzian electrodynamics) into the possibility of viewing geometry, not as uniquely forced upon us by either reason or experience, but rather as resting on a free choice, a convention of our own. Einstein, in the special theory of relativity, then applies this insight to the concepts of time and simultaneity.[27]

Science, if it is to continue to progress through revolutions, therefore needs a source of new ideas, alternative programs, and expanded possibilities that is not itself scientific in the same sense – that does not, as do the sciences themselves, operate within a generally agreed upon framework of taken for granted rules. For what is needed here is precisely the creation and stimulation of new frameworks or paradigms, together with what we might call meta-frameworks or meta-paradigms – new conceptions of what a coherent rational understanding of nature might amount to – capable of motivating and sustaining the revolutionary transition to a new first-level or scientific paradigm. Philosophy, throughout its close association with the sciences, has functioned in precisely this way. (And I would imagine that parallel points can be made, *mutatis mutandis*, in the case of some of the other humanities.)[28]

---

27. For discussion of Einstein's reading of Poincaré in 1905 see Miller (1981, chapter 2). This example is further discussed in Part Two, section 4 below.
28. See, for example, Edgerton (1991) for very interesting suggestions about the relationship between the development of renaissance linear perspective and the (seventeenth century) scientific revolution.

From this point of view, it is folly for philosophy to attempt to incorporate itself into the sciences (as a branch of psychology, say, or mathematical logic), for its peculiar role is precisely to articulate and stimulate new possibilities, at the meta-scientific level, as it were, and it cannot, on pain of entirely relinquishing this role, itself assume the position of a normal science. For the same reason, it is also folly for philosophy to attempt to become "scientific," in the sense of finally leaving behind the traditional conflict of opposing schools for a new stable consensus on generally agreed upon rules of inquiry. We never know in advance what new paradigms (and philosophical meta-paradigms) might be needed at a given moment of revolutionary science, and so, in philosophy (and, *mutatis mutandis*, also in the other humanities), it is always to our advantage to let a thousand flowers bloom. Finally, it is folly as well for philosophy (and for the other humanities) to regret this lack of scientific status, and, even worse, to seek compensation by attempting to strip away such status from the sciences themselves. We should rather rejoice, along with the sciences, in our fundamentally distinct, yet mutually complementary contributions to the total ongoing dialectic of human knowledge.

II

Historical Perspectives on the Stratification of Knowledge

In Lecture I we observed that Kant's idea of a "transcendental philosophy" is the historical source for the differentiation between philosophy and the special sciences that has dominated the modern period. Philosophy, for Kant, no longer had a characteristic first-level subject matter or object of its own – such as the immaterial and invisible part of the universe, God and the soul, for example – but functioned rather at a second or meta-level to describe what Kant called the conditions of possibility of first-level scientific knowledge. And the first-level scientific knowledge Kant had primarily in mind, as we also observed, was the Newtonian mathematical physics and astronomy that represented the paradigm of knowledge of nature throughout the eighteenth century. Kant frames his "transcendental" inquiry, accordingly, by the two questions "How is pure mathematics possible?" and "How is pure natural science possible?" – where the first concerns above all, the possibility of Euclidean geometry (which is of course taken to be the geometry of physical space in Newtonian physics), and the second concerns the possibility of fundamental laws of Newtonian mechanics such as conservation of mass, inertia, and equality of action and reaction.

These two questions as to the possibility of pure mathematics and pure natural science are two particular specifications of what Kant calls "the general problem of pure reason," namely, "How are synthetic a priori judgements possible?"[29] This problem has been bequeathed to us by David Hume, Kant

---

29. This "general problem of pure reason," along with its two more specific sub-problems, is formulated in section VI of the Introduction to the *Critique of Pure Reason* at B19–24. Sections V and VI, which culminate in the three questions "How is pure mathematics possible?", "How is pure natural science possible?", and "How is metaphysics as a science possible?", are added to the second (1787) edition of the *Critique* and clearly follow the structure of the 1783 *Prolegomena to Any Future Metaphysics*, which was intended to clarify the first (1781) edition. This way of framing the general problem of pure reason also clearly reflects the increasing emphasis on the question of

suggests, who first emphasized, in opposition to traditional rationalism (such as that of Descartes and Leibniz), that no substantive knowledge of the empirical spatio-temporal world could possibly be provided by purely logical or analytic judgements – by what Hume termed "relations of ideas." For it is never simply self-contradictory to deny that any such spatio-temporal relation actually obtains. (Hume had the causal relation primarily in mind, of course.) Kant cannot follow Hume, however, in a general skepticism about rational or a priori knowledge of such relations. For Kant takes it as self-evident that certain basic parts of Newtonian physics – Euclidean geometry and the fundamental laws of mechanics – obviously have a special a priori status.[30] These parts of Newtonian physics are not derived empirically, by any kind of inductive extrapolation from our experience of nature. On the contrary, the possibility of any such inductive extrapolation (of Newton's argument from "phenomena" to the law of universal gravitation, for example) *presupposes* that the more fundamental laws of geometry and mechanics are already in place (Newton's argument culminating in universal gravitation, for example, cannot even begin without geometry and mechanics). The general problem of pure reason is therefore solved, for Kant, by the idea that, whereas synthetic a priori knowledge (typified by geometry and mechanics) is not, as Hume saw, straightforwardly logically or analytically true, it nevertheless functions as the presupposition or condition of possibility of all properly empirical knowledge.

It must be conceded, I believe, that Kant has the upper hand in this eighteenth century philosophical debate with Hume. For, in this context, there are

---

pure natural science found in the *Metaphysical Foundations of Natural Science* (1786). For an extended discussion of Kant's theory of pure natural science and its relation to Newtonian physics see Friedman (1992, especially chapters 3 and 4).

30. After introducing the two problems as to the possibility of pure mathematics and pure natural science Kant remarks (B20): "Since these sciences actually exist, it can now be appropriately asked *how* they are possible, for that they must be possible is proved by their actuality.*" The footnote then continues: "Some might still have doubts about pure natural science. But one need only consider the various propositions that occur at the beginning of proper (empirical) physics, such as the permanence of the same quantity of matter, inertia, and the equality of action and reaction, etc., to be soon convinced that they constitute a *physica pura* (or *rationalis*), which well deserves to be separately articulated as an independent science in its entire extent, whether narrow or wide." Immediately before this passage Kant explicitly considers Hume's skeptical attitude towards the principle of causality and remarks (ibid.): "[Hume] would have never fallen upon this [skeptical] assertion if he had entertained our problem in its [full] generality; for he would have realized then that, according to his argument, there could also be no pure mathematics, for this certainly contains synthetic a priori propositions – from which assertion his good sense would have surely protected him." Compare also the remarks about Locke and Hume at B127–8.

simply no conceivable alternatives to Euclidean geometry and Newtonian mechanics. If one wants an empirical science of nature at all in the eighteenth century then one simply has no option, as Kant incisively argues, but to pre-suppose Euclidean geometry and the laws of mechanics as given, on the basis of which one can then proceed to elaborate empirical laws of nature such as the law of universal gravitation.[31] In this context, although Hume, like every-one else, of course takes universal gravitation as the paradigm of natural science, the rather meager picture of empirical inquiry that emerges from Hume's account cannot possibly do justice to Newton's actual argument – to the startling conclusion, in particular, that the law of gravitation holds with absolutely strict universality (so that, for example, there is an instantaneous gravitational interaction at this moment between the tip of my finger and the planet Pluto). By contrast, Kant, as I suggested in Lecture I, has a deep and penetrating account of the Newtonian argument, which, in particular, clari-fies fundamental conceptual difficulties surrounding concepts like absolute space and action at a distance.

But the beginning of the nineteenth century brought with it a profound challenge to the Kantian conception of a priori knowledge from a quite unex-pected direction – the development of alternative geometries differing essentially from Euclid's classic system. If one of these non-Euclidean geome-tries were true of physical space then Euclid's geometry would not even be true in fact, much less synthetic a priori true. Moreover, even if physical space is in fact Euclidean, we still have the problem – the entirely new problem – of explaining how we know that Euclidean geometry, *as opposed to one of its non-Euclidean alternatives*, is in fact true. Finally, since we have now succeeded, at least apparently, in conceiving the possibility that space might be other than Euclidean, Kant's idea that Euclidean geometry is built into the fundamental capacities of the human mind (into what Kant called our pure intuition of space) appears to be simply false.

This challenge to the Kantian synthetic a priori figured prominently in the thinking of the scientific philosophers of the nineteenth century, in the thought of Hermann von Helmholtz and Henri Poincaré, in particular. And it was the work of these nineteenth century thinkers, as I observed in Lecture I, that eventuated in the development of Einstein's theory of relativity in the early years of the twentieth century, wherein a new non-Euclidean and non-

---

31. As is well known, Hume explicitly denies that the laws of mechanics are a priori in the *Enquiry*, Section IV, Part II, and he even extends his skepticism to parts of geometry in the *Trea-tise*, Book I, Part II, Section IV, § III (contrary to Kant's sanguine expectations expressed in note 30 above).

Newtonian geometry and mechanics is actually applied to our experience of nature – in the form specifically, of a radically new conceptualization of gravitation in which gravitational force acting instantaneously at a distance no longer appears. In Einstein's new theory of gravitation, the general theory of relativity, the curvature of space (more precisely, of space-time) replaces gravitational force: bodies do not deviate from straight lines (in Euclidean space) due to the action of gravitational force, but instead follow straightest possible lines or geodesics of a new type of non-Euclidean geometry. So, according to this theory, Euclidean geometry in fact fails to hold of physical space (of the space of our solar system, for example).

What, then, is the fate of the Kantian conception of a priori knowledge? It is fashionable at the present time to argue – often on the basis of precisely the nineteenth and early twentieth century developments in the foundations of geometry and mathematical physics just reviewed – that there are no a priori elements at all in our scientific knowledge. If Euclidean geometry, at one time the very model of rational or a priori knowledge of nature, can be empirically revised, so the argument goes, then everything is in principle empirically revisable. Our reasons for adopting one or another system of geometry or mechanics (or, indeed, of mathematics more generally or of logic) are at bottom of the very same kind as the purely empirical considerations that support any other part of our total theory of nature. Our system of knowledge, in W. V. Quine's well-known figure, should be viewed holistically as a vast web of interconnected beliefs on which experience or sensory input impinges only along the periphery. When faced with a "recalcitrant experience" standing in conflict with our overall system of beliefs we then have a choice of where to make revisions. These can be made relatively close to the periphery of the system (in which case we make a change in a relatively low-level part of natural science), but they can also – when the conflict is particularly acute and persistent, for example – affect the most abstract and general parts of science, including even the truths of logic and mathematics, lying at the center of our system of beliefs. To be sure, such high-level beliefs at the center of our system are relatively entrenched, in that we are relatively reluctant to revise them or to give them up (as we once were in the case of Euclidean geometry, for example). Nevertheless, and this is the crucial point, absolutely none of our beliefs is forever "immune to revision" in light of experience:

> The totality of our so-called knowledge or beliefs, from the
> most casual matters of geography and history to the profoundest
> laws of atomic physics or even of pure mathematics and logic, is

a man-made fabric which impinges on experience only along the edges. Or, to change the figure, total science is like a field of force whose boundary conditions are experience. A conflict with experience at the periphery occasions readjustments in the interior of the field. … But the total field is so underdetermined by its boundary conditions, experience, that there is much latitude of choice as to what statements to reëvaluate in the light of any single contrary experience. …

If this view is right … it becomes folly to seek a boundary between synthetic statements, which hold contingently on experience, and analytic statements, which hold come what may. Any statement can be held true come what may, if we make drastic enough adjustments elsewhere in the system. Even a statement very close to the periphery can be held true in the face of recalcitrant experience by pleading hallucination or by amending certain statements of the kind called logical laws. Conversely, by the same token, no statement is immune to revision. Revision even of the logical law of the excluded middle has been proposed as a means of simplifying quantum mechanics; and what difference is there in principle between such a shift and the shift whereby Kepler superseded Ptolemy, or Einstein Newton, or Darwin Aristotle?[32]

As the last sentence makes clear, examples of revolutionary transitions in our scientific knowledge, and, in particular, that of the Einsteinian revolution in geometry and mechanics, constitute a very important part of the motivations for this view.

Yet it is important to see that such a strongly anti-apriorist conception of scientific knowledge was by no means prevalent during the late nineteenth and early twentieth century – during the very period, that is, when the great revolutions in geometry and mechanics we now associate with Einstein were actually taking place. In the case of Helmholtz, for example, although it is true that he viewed the choice between Euclidean and non-Euclidean geometries as an empirical one, he also suggested that the more general structure of space common to both Euclidean and non-Euclidean systems (that of constant curvature or what Helmholtz called "free mobility") was a necessary

---

32. From the first two paragraphs of section 6 – entitled "Empiricism without the dogmas" – of Quine (1951/53, pp. 42–3).

presupposition of all spatial measurement and thus a "transcendental" form of our spatial intuition in the sense of Kant. And, partly on this basis, Poincaré went even further. Although no particular geometry – neither Euclidean nor non-Euclidean – is an a priori condition of our spatial intuition, it does not follow that the choice between such geometries is an empirical one. For there remains an irreducible gulf between our crude and approximate sensory experience and our precise mathematical descriptions of nature. Establishing one or another system of geometry, Poincaré argued, therefore requires a free choice, a *convention* of our own – based, in the end, on the greater mathematical simplicity of the Euclidean system (Poincaré died before the general theory of relativity was created).[33]

Nor was such a strongly anti-apriorist conception of scientific knowledge adopted by the first scientific thinkers enthusiastically to embrace Einstein's new theory. These thinkers, the logical empiricists, of course rejected the synthetic a priori in Kant's original form. They rejected the idea of absolutely fixed and unrevisable a priori principles built, once and for all, into our fundamental cognitive capacities. In place of an holistic empiricism, however, they instead adopted variants of Poincaré's conventionalism. Such conventionalism, as we noted in Lecture I, was central to Moritz Schlick's interpretation of relativity theory, for example, as well as to Rudolf Carnap's closely related discussion. But perhaps the clearest articulation of the logical empiricists's new view of a priori principles was provided by Hans Reichenbach in his first book, *The Theory of Relativity and A Priori Knowledge*, published in 1920.[34] Reichenbach distinguishes two meanings of the Kantian a priori: necessary and unrevisable, fixed for all time, on the one hand, and "constitutive of the concept of the object of [scientific] knowledge," on the other. Reichenbach argues, on this basis, that the great lesson of the theory of relativity is that the former meaning must be dropped while the latter must be retained. Relativity theory involves a priori constitutive principles as necessary presuppositions of its properly empirical claims, just as much as did Newtonian physics, but these principles have essentially changed in the transition from the latter theory to the former: whereas Euclidean geometry is indeed constitutively a priori in the context of Newtonian physics, for example, only *infinitesimally* Euclidean geometry – consistent with all possible values of the

---

33. For extended discussion of Helmholtz and Poincaré – including their relationships to both Kant and the logical empiricists – see Friedman (1999, chapter 4), (1997b), (2000a).

34. Reichenbach (1920/65). The distinction between the two meanings of the Kantian a priori described in the next sentence occurs in chapter 5, entitled "Two meanings of the a priori and Kant's implicit presupposition."

curvature – is constitutively a priori in the context of general relativity. What we end up with, in this tradition, is thus a relativized and dynamical conception of a priori mathematical-physical principles, which change and develop along with the development of the mathematical and physical sciences themselves, but which nevertheless retain the characteristically Kantian constitutive function of making the empirical natural knowledge thereby structured and framed by such principles first possible.[35]

Rudolf Carnap's philosophy of formal languages or linguistic frameworks, first developed in his *Logical Syntax of Language* in 1934 and then reemphasized, in the context of later debates with Quine, in his paper "Empiricism, Semantics, and Ontology," published in 1950, was the most mature expression of the logical empiricists's new view.[36] All standards of "correctness," "validity," and "truth," according to Carnap, are relative to the logical rules or principles definitive of one or another formal language or linguistic framework. The rules of classical logic and mathematics, for example, are definitive of certain logical calculi or linguistic frameworks, while the rules of intuitionistic logic and mathematics (wherein the law of excluded middle is no longer universally valid) are definitive of others. Since standards of "validity" and "correctness" are thus relative to the choice of linguistic framework, it makes no sense to ask whether any such choice of framework is itself "valid" or "correct." For the logical rules relative to which alone these notions can first be well defined are not yet in place. Such rules are *constitutive* of the concepts of "validity" and "correctness" – relative to one or another choice of linguistic framework, of course – and are in this sense a priori rather than empirical.

This Carnapian philosophy of linguistic frameworks rests on two closely related distinctions. The first is the distinction between formal or *analytic* sentences of a given framework and empirical or *synthetic* sentences – or, as Carnap puts it in *Logical Syntax*, between *logical rules* ("L-rules") of a linguis-

---

35. For discussion of Kant's characteristic conception of the constitutive function of a priori principles see De Pierris (1993).

36. As we observed in note 18 above, Carnap first explicitly introduces the notion of linguistic framework and the distinction between internal and external questions in this latter paper. Carnap (1950/56, footnote 5, p. 215) refers to Quine (1948/53) and remarks that "Quine does not acknowledge the distinction [between internal and external questions] because according to his general conception there are no sharp boundary lines between logical and factual truth, between questions of meaning and questions of fact, between the acceptance of a language structure and the acceptance of an assertion formulated in the language." It is clear, then, that the analyticity debate – which originated in discussions at Harvard in the years 1939–41 involving Carnap, Quine, and Alfred Tarski – constitutes the immediate philosophical background for this divergence. For discussion of the analyticity debate see the Introduction to Creath (1990).

tic framework and *physical rules* ("P-rules"). The L-rules include laws of logic and mathematics (and also, at least in spaces of constant curvature, laws of geometry), whereas the P-rules include empirical laws standardly so-called such as Maxwell's equations of electromagnetism. In this way, Carnap's distinction between L-rules and P-rules closely parallels Reichenbach's distinction, developed in his 1920 book on relativity theory and a priori knowledge, between "axioms of coordination" (constitutive principles) and "axioms of connection" (properly empirical laws).[37] Carnap's differentiation between logical and physical rules of a framework (analytic and synthetic sentences) then induces a second fundamental distinction between *internal* and *external* questions. Internal questions are decided within an already adopted framework, in accordance with the logical rules of the framework in question. External questions, by contrast, concern precisely the question of which linguistic framework – and therefore, which logical rules – to adopt in the first place. And, since no logical rules are as yet in place, external questions, unlike internal questions, are not rationally decidable in the same sense (they are not strictly speaking capable of answers that are true or false, for example). Such questions can only be decided conventionally on the basis of broadly pragmatic considerations of convenience or suitability for one or another purpose. An overriding desire for security against the possibility of contradiction, for example, may prompt the choice of the weaker rules of intuitionistic logic and mathematics, whereas an interest in ease of physical application may prompt the choice of the stronger rules of classical logic and mathematics. (As we indicated in Lecture I, Carnap uses the distinction between internal and external questions to dissolve and diffuse the traditional problems of philosophy – all of which are now understood in terms of *practical proposals* to formulate the total language of science using the rules of one or another linguistic framework, rather than as genuine *theoretical questions* that would be rationally decidable given some antecedent choice of such rules.)

Now it was precisely this Carnapian philosophy of linguistic frameworks that formed the background and foil for Quine's articulation of a radically opposed form of epistemological holism according to which no fundamental distinction between a priori and a posteriori, logical and factual, analytic and synthetic can in fact be drawn. Indeed, it was in Quine's 1951 paper "Two Dogmas of Empiricism" (cited in note 32 above), where his challenge to the analytic/synthetic distinction was first made widely known, that the holistic

---

37. For discussion of the relationship between Carnap's philosophy of linguistic frameworks and Reichenbach's 1920 book see Friedman (1999, chapter 3).

figure of knowledge as a vast web of interconnected beliefs also first appeared. And it is important to see here that it is Quine's attack on the analytic/synthetic distinction, and not simply the idea that no belief whatsoever is forever immune to revision, that is basic to Quine's new form of epistemological holism. For Carnap's philosophy of linguistic frameworks is predicated on the idea that logical or analytic principles, just as much as empirical or synthetic principles, can be revised in the progress of empirical science.[38] Indeed, as we have seen, Reichenbach's initial formulation of this new view of constitutive a priori principles was developed precisely to accommodate the revolutionary changes in the geometrical and mechanical framework of physical theory wrought by Einstein's development of the theory of relativity. The difference between Quine and Carnap is rather that the latter persists in drawing a sharp distinction between changes of language or linguistic framework, in which constitutive principles definitive of the very notions of "validity" and "correctness" are revised, and changes in ordinary empirical statements formulated against the background of such an already present constitutive framework. And this distinction, for Carnap, ultimately rests on the difference between analytic statements depending solely on the meanings of the relevant terms and synthetic statements expressing contentful assertions about the empirical world.

Quine's attack on Carnap's version of the analytic/synthetic distinction – and thus on Carnap's version of the distinction between a priori and empirical principles – is now widely accepted, and I have no desire to defend Carnap's particular way of articulating this distinction here. I do want to question, however, whether Quinean epistemological holism is really our only option, and whether, in particular, such epistemological holism represents our best way of coming to terms with the revolutionary changes in the historical development of science that are now often taken to support it.

Quine's epistemological holism pictures our total system of science as a vast web or conjunction of beliefs which face the "tribunal of experience" as a corporate body. Quine grants that some beliefs, such as those of logic and arithmetic, for example, are relatively central, whereas others, such as those of biology, say, are relatively peripheral. But this means only that the former

---

38. Carnap explicitly embraces this much of epistemological holism (based on the ideas of Poincaré and Pierre Duhem) in section 82 of *Logical Syntax*. Quine is therefore extremely misleading when he (in the above-cited passage from section 6 of "Two Dogmas") simply equates analyticity with unrevisability. He is similarly misleading in section 5 (1951/53, p. 41) when he asserts that the "dogma of reductionism" (i.e., the denial of Duhemian holism) is "at root identical" with the dogma of analyticity.

beliefs are less likely to be revised in case of a "recalcitrant experience" at the periphery, whereas the latter are more likely to be revised. Otherwise, from an epistemological point of view, there is simply no relevant distinction to be made here:

> Suppose an experiment has yielded a result contrary to a theory currently held in some natural science. The theory comprises a whole bundle of conjoint hypotheses, or is resoluble into such a bundle. The most that the experiment shows is that at least one of these hypotheses is false; it does not show which. It is only the theory as a whole, and not any of the hypotheses, that admits of evidence or counter-evidence in observation and experiment.
>
> And how wide is a theory? No part of science is quite isolated from the rest. Parts as disparate as you please may be expected to share laws of logic and arithmetic, anyway, and to share various common-sense generalities about bodies in motion. Legalistically, one could claim that evidence counts always for or against the total system, however loose-knit, of science. Evidence against the system is not evidence against any one sentence rather than another, but can be acted on rather by any of various adjustments.
>
> Thus suppose from a combined dozen of our theoretical beliefs a scientist derives a prediction in molecular biology, and the prediction fails. He is apt to scrutinize for possible revision only the half dozen beliefs that belonged to molecular biology rather than tamper with the more general half dozen having to do with logic and arithmetic and the gross behavior of bodies. This is a reasonable strategy – a maxim of minimum mutilation. But an effect of it is that the portion of theory to which the discovered failure of prediction is relevant seems narrower than it otherwise might.[39]

Strictly speaking, then, empirical evidence – either for or against – spreads over all the elements of the vast conjunction that is our total system of science, wherein all elements whatsoever equally face the "tribunal of

---

39. Quine (1970, pp. 5, 7).

experience." And it is in this precise sense, for Quine, that all beliefs whatsoever, including those of logic and mathematics, are equally empirical.

But can this beguiling form of epistemological holism really do justice to the revolutionary developments within both mathematics and natural science that have led up to it? Let us first consider the Newtonian revolution that produced the beginnings of mathematical physics as we know it – the very revolution, as we have seen, Kant's original conception of synthetic a priori knowledge was intended to address. In constructing his mathematical physics Newton created, virtually simultaneously, three revolutionary advances: a new form of mathematics, the calculus, for dealing with infinite limiting processes and instantaneous rates of change; new conceptions of force and quantity of matter embodied and encapsulated in his three laws of motion; and a new universal law of nature, the law of universal gravitation. Each of these three advances was revolutionary in itself, and all were introduced by Newton in the context of the same scientific problem: that of developing a single mathematical theory of motion capable of giving a unified account of both terrestrial and celestial phenomena. Since all of these advances were thus inspired, in the end, by the same empirical problem, and since they together amounted to the first known solution to this problem, Quine's holistic picture appears so far correct. All elements in this web or conjunction of scientific knowledge – mathematics, mechanics, gravitational physics – appear equally to face the "tribunal of experience" together.

Nevertheless, there are fundamental asymmetries in the way in which the different elements of this Newtonian synthesis actually function. Consider, for example, the relationship between the mathematics of the calculus and the Newtonian formulation of the laws of mechanics. Newton's second law of motion (in only slightly anachronistic form) says that force equals mass times acceleration, where acceleration is the instantaneous rate of change of velocity (itself the instantaneous rate of change of position); so without the mathematics of the calculus (the mathematics of infinite limiting processes and instantaneous rates of change) this second law of motion could not even be formulated or written down, let alone function to describe empirical phenomena.[40] The combination of calculus plus the laws of motion is not

---

40. As is well known, Newton did not himself formulate the second law as we would formulate it today: rather then a continuously acting force producing an acceleration or change of momentum in a given (or unit) time, Newton employed an instantaneous "impulsive" force producing a given (finite) change of momentum. He then moved from discrete impulsive forces to continuously acting (accelerative) forces by a geometrical limiting process. For discussion see Cohen (1999, sections 5.3 and 5.4).

happily viewed, therefore, as a conjunction of elements symmetrically contributing to a single total result. For one element of a conjunction can always be dropped while the second remains with its meaning and truth-value intact. In this case, however, the mathematics of the calculus does not function simply as one more element in a larger conjunction, but rather as a necessary presupposition without which the rest of the putative conjunction has no meaning or truth-value at all. The mathematical part of Newton's theory therefore supplies elements of the language or conceptual framework, we might say, within which the rest of the theory is then formulated.

An analogous (if also more complex and subtle) point holds with respect to the relationship between Newton's mechanics and gravitational physics. The law of universal gravitation says that there is a force of attraction or approach, directly proportional to the product of the two masses and inversely proportional to the square of the distance between them, between any two pieces of matter in the universe. Any two pieces of matter therefore experience accelerations towards one another in accordance with the very same law. But relative to what frame of reference are the accelerations in question defined? Since these accelerations are, by hypothesis, universal, they affect every piece of matter in the universe; so no particular material body can be taken as actually at rest in this frame, and thus the motions in question are not motions relative to any particular material body. Newton himself understood these motions as defined relative to absolute space, but we now understand Newtonian theory differently. The privileged frame of reference in which the law of gravitation is defined is what we now call an *inertial frame*, where an inertial frame of reference is simply one in which the Newtonian laws of motion hold (the center of mass frame of the solar system, for example, is a very close approximation to such a frame). It follows that without the Newtonian laws of mechanics the law of universal gravitation would not even make empirical sense, let alone give a correct account of the empirical phenomena. For the concept of universal acceleration that figures essentially in this law would then have no empirical meaning or application: we would simply have no idea what the relevant frame of reference might be in relation to which such accelerations are defined.[41]

---

41. One way to see this is to consider the choice between Copernican-Keplerian and Ptolemaic-Tychonic models of the solar system before Newton's *Principia*. In this case one simply has two kinematically equivalent descriptions with no dynamical grounds for choosing between them: the former system holds relative to the sun and the latter relative to the earth. To give meaning to the assertion that the earth really revolves around the sun rather than the other way around (more precisely, around the center of gravity of the solar system) one needs the new dynamical conception of motion expressed in Newton's laws.

Once again, Newton's mechanics and gravitational physics are not happily viewed as symmetrically functioning elements of a larger conjunction: the former is rather a necessary part of the language or conceptual framework within which alone the latter makes empirical sense.

Now it was just such an analysis of Newton's mathematical physics, we have suggested, that inspired Kant's original conception of synthetic a priori knowledge. It was for precisely these reasons that Kant took the mathematical part of Newton's theory (what we would now call calculus on a Euclidean space), as well as the laws of motion or mechanics, to have a fundamentally different status than the empirical parts of the theory such as the law of gravitation: the former, according to Kant, constitute the necessary presuppositions or conditions of possibility of the latter.[42] Moreover, since, in Kant's day, the Newtonian system was the only mathematical physics the world had ever seen (a condition of innocence it is very hard for us today imaginatively to recapture), Kant also took these necessary conditions of the possibility of Newtonian physics to be absolutely fixed conditions of all future empirical science in general. The development of non-Euclidean geometries and Einsteinian relativity theory of course then demolishes this second aspect of Kant's original doctrine. But does it also undermine the first aspect – the idea of necessary presuppositions constituting the conditions of possibility of the properly empirical parts of a scientific theory? Does it also support Quinean holism against the modified, relativized and dynamical conception of a priori constitutive principles developed by the logical empiricists?

I think not. The general theory of relativity, like Newton's original theory of gravitation, can be seen as the outcome of three revolutionary advances: the development of a new field of mathematics, tensor calculus or the general theory of manifolds (originally developed by Bernhard Riemann in the latter part of the nineteenth century); Einstein's principle of equivalence, which identifies gravitational effects with the inertial effects formerly associated with Newton's laws of motion; and Einstein's equations for the gravitational field, which describe how the curvature of space-time is modified by the presence of matter and energy so as to direct gravitationally affected bodies along the straightest possible paths or geodesics of the space-time geometry in the

---

42. I have argued in detail (in the reference cited in notes 9, 10, and 29 above) that Kant's theory of pure natural science is based on an analysis of the relationship between the Newtonian laws of motion and the law of universal gravitation closely analogous to the one presented immediately above. The difference is that Kant does not have the concept of inertial frame and instead views the Newtonian laws of motion (together with other fundamental principles Kant takes to be a priori) as defining a convergent sequence of ever better approximations to a single privileged frame of reference (a counterpart of absolute space) at rest at the center of gravity of all matter.

relevant region. Once again, each of these three advances was revolutionary in itself, and all three were marshalled together by Einstein to solve a single empirical problem: that of developing a new description of gravitation consistent with the special theory of relativity (which is itself incompatible with the instantaneous action at a distance characteristic of Newtonian theory) and also capable, it was hoped, of solving well-known anomalies in Newtonian theory such as that involving the perihelion of Mercury. And the three advances together, as marshalled and synthesized by Einstein, in fact succeeded in solving this empirical problem for the first time.

It does not follow, however, that the three advances in question can be happily viewed as symmetrically functioning elements of a single conjunction, which then equally face the "tribunal of experience" together when confronted with the anomaly in the perihelion of Mercury, for example. Consider first the relationship between the first two advances. The principle of equivalence depicts the space-time trajectories of bodies affected only by gravitation as geodesics or straightest possible paths in a variably curved space-time geometry, just as the Newtonian laws of motion, when viewed from this same space-time perspective, depict the trajectories of bodies affected by no forces at all as geodesics or straightest possible paths in a flat or Euclidean space-time geometry.[43] But the whole notion of a variably curved geometry itself only makes sense in the context of the revolutionary new mathematics of the general theory of manifolds recently created by Riemann. In the context of the mathematics available in the seventeenth and eighteenth centuries, by contrast, the idea of a variably curved space-time geometry could not even be formulated or written down, let alone function to describe empirical phenomena. And, once again, a closely analogous, but also more subtle point holds for the relationship between the second and third advances. Einstein's field equations describe the variations in curvature of space-time geometry as a function of the distribution of mass and energy. Such a variably curved space-time structure would have no empirical meaning or application, however, if we had not first singled out some empirically given phenomena as counterparts of its fundamental geometrical notions – here the notion of geodesic or straightest possible path. The principle of equivalence does precisely this, however, and without this principle the intricate space-time geometry described by Einstein's field equations would not

---

43. This is what the classical law of inertia looks like from a modern, four dimensional point of view. For discussion see Earman and Friedman (1973), Friedman (1983, section III.7). This point of view was first introduced into recent philosophical discussions by Stein (1967).

even be empirically false, but rather an empty mathematical formalism with no empirical application at all.[44] Just as in the case of Newtonian gravitation theory, therefore, the three advances together comprising Einstein's revolutionary theory should not be viewed as symmetrically functioning elements of a larger conjunction: the first two function rather as necessary parts of the language or conceptual framework within which alone the third makes both mathematical and empirical sense.

It will not do, in either of our two examples, to view what I am calling the constitutively a priori parts of our scientific theories as simply relatively fixed or entrenched elements of science in the sense of Quine, as particularly well established beliefs which a reasonable scientific conservatism takes to be relatively difficult to revise (following what Quine, in the quotation from *Philosophy of Logic* presented above, calls a "maxim of minimum mutilation"). When Newton formulated his theory of gravitation, for example, the mathematical part of his theory, the new calculus, was still quite controversial – to such an extent, in fact, that Newton disguised his use of it in the *Principia* in favor of traditional synthetic geometry.[45] Nor were Newton's three laws of motion any better entrenched, at the time, than the law of universal gravitation.[46] Similarly, in the case of Einstein's general theory of relativity, neither the mathe-

---

44. For an analysis of the principle of equivalence along these lines, including illuminating comparisons with Reichenbach's conception of the need for "coordinating definitions" in physical geometry, see DiSalle (1995). This matter is further discussed in detail in Part Two below, especially sections 1 and 2.

45. As indicated in note 40 above, Newton used geometrical limiting arguments where we would now use the (algebraically formulated) calculus. Instead of setting force proportional to the time rate of change of momentum, for example, Newton lets an infinite number of instantaneous "impulsive" forces approach a continuously acting force in the limit. For a very detailed discussion of the mathematics of the *Principia* see Guicciardini (1999).

46. Newton himself portrays the laws of motion as already familiar and accepted, as natural generalizations of the work of such predecessors as Galileo and Huygens. In one sense, this is perfectly correct – and, indeed, the laws of motion are perhaps most illuminatingly viewed as generalizations of the conservation of momentum principle contained in the laws of impact, extended to include continuously acting forces on the model of Galileo's treatment of uniform acceleration. (Here I am indebted to discussions with Domenico Bertoloni Meli.) However, the way Newton actually deploys these laws – especially the third law, the equality of action and reaction – essentially involves an application to action-at-a distance forces which, as such, would be entirely unacceptable from the point of view of the then dominant mechanical philosophy. In this sense, Newton's laws of motion constitute a quite radical *transformation* of the previous tradition. This issue is closely connected with the mathematical issues discussed in notes 40 and 45 above. For, as Cohen (1999, section 5.3) suggests, it appears that Newton begins with discrete "impulsive" forces precisely to make his radically new application to (continuous) action-at-a-distance forces more palatable to proponents of the mechanical philosophy.

matical theory of manifolds nor the principle of equivalence was a well-
entrenched part of main-stream mathematics or mathematical physics.[47] And
this is one of the central reasons, in fact, that Einstein's theory was so profoundly revolutionary. More generally, then, since we are dealing with deep
conceptual revolutions in both mathematics and mathematical physics in
both cases, entrenchment and relative resistance to revision are not appropriate distinguishing features at all. What characterizes the distinguished elements of our theories is rather their special *constitutive function*: the function
of making the precise mathematical formulation and empirical application of
the theories in question first possible. In this sense, the relativized and dynamical conception of the a priori developed by the logical empiricists appears to
describe these conceptual revolutions far better than does Quinean holism.
This is not at all surprising, in the end, for this new conception of the constitutive a priori was inspired, above all, by just these conceptual revolutions.

Where, then, does Quinean epistemological holism derive its force? Only
as a reaction, I want to suggest, to the particular attempt precisely to characterize the notion of constitutive a priori principles found in Carnap's work.
Carnap's theory of formal languages or linguistic frameworks was to be developed, as we saw in Lecture I, within what Carnap called *Wissenschaftslogik*
and, as such, within mathematical logic. This, in fact, was how philosophy
was to become scientific for Carnap (as a part of *formal* or mathematical science). In particular, then, the fundamental distinction, in the context of any
given formal language or linguistic framework, between logical and empirical
rules, analytic and synthetic sentences, was itself supposed to be a purely formal or logical distinction. But one of the main pillars of Quine's attack on the
analytic/synthetic distinction is simply that, from a purely formal or logical
point of view, all sentences derivable within a given formal system are so far
completely on a par – so that, more specifically, Carnap's attempt further to
characterize some subset of derivable sentences as analytic ultimately
amounts to nothing more than an otherwise arbitrary label.[48] And, more
generally, Quine sees no reason to restrict the methods of scientific philoso-

---

47. As I observe in Part Two below (note 12), there were applications of Riemannian geometry in
late nineteenth century Hamiltonian mechanics. However, the Riemannian theory of manifolds
was not extensively developed, even within pure mathematics, until after Einstein's work on the
general theory of relativity.
48. For the technical problems afflicting Carnap's attempt to provide a general definition of analyticity (in what he calls "general syntax") in *Logical Syntax* see Friedman (1999, Part Three). For
Quine, this Carnapian failure to provide a general definition then means that any attempt to
explain analyticity on a language-by-language basis must be seen as essentially arbitrary.

phy to purely formal or logical ones. In particular, the resources of behavioristic psychology (to which Carnap himself had also earlier appealed) allow us to enrich the descriptive capabilities of Carnapian epistemology or *Wissenschaftslogik* with further scientific resources. Within this enlarged point of view, the point of view Quine calls epistemology naturalized, we then see that all that remains of the a priori is precisely relative centrality or entrenchment: some sentences are more likely than others to be revised under prompting by sensory stimulation. So, from a strictly scientific point of view, Quine concludes, we must accept epistemological holism.[49]

My own conclusion, however, is quite different. Quine is correct that pure formal logic is insufficient to characterize the relativized and dynamical, yet still constitutive notion of a priori principles Carnap was aiming at. Quine is also correct that behavioristic psychology, the study of the input-output relations of the human organism, is also quite insufficient for this purpose – for all we have, from this point of view, is the notion of relative entrenchment or resistance to revision. Yet, as we have seen, careful attention to the actual historical development of science, and, in particular, to the profound conceptual revolutions that have in fact led to our current philosophical predicament, shows that relativized a priori principles of just the kind Carnap was aiming at are central to our scientific theories. Although Carnap may have failed in giving a precise logical characterization or explication of such principles, it does not follow that the *phenomenon* he was attempting to characterize does not exist. On the contrary, everything we know about the history of science, I want to suggest, indicates that precisely this phenomenon is an absolutely fundamental feature of science as we know it – and a fundamental feature, in particular of the great scientific revolutions that have eventually led, in our time, to the Carnap-Quine debate.

I take this last idea to be strongly confirmed by the circumstance that in Thomas Kuhn's theory of the nature and character of scientific revolutions we find an informal counterpart, in effect, of the relativized conception of constitutive a priori principles developed by the logical empiricists. Thus, Kuhn's central distinction between change of paradigm or revolutionary science, on the one side, and normal science, on the other, closely parallels the Carnapian distinction between change of language or linguistic framework and rule-governed operations carried out within such a framework. Just as, for Carnap, the logical rules of a linguistic framework are definitive or constitutive of

---

49. Quine's epistemology naturalized thus amounts to a behavioristic version of the standpoint earlier represented by Helmholtz: philosophy as a part of *empirical* science, viz., psychology.

the notion of "correctness" or "validity" relative to this framework, so a particular paradigm governing a given state or episode of normal science, for Kuhn, yields generally agreed upon (although perhaps only tacit) rules definitive or constitutive of what counts as a "valid" or "correct" solution to a problem within this state of normal science. Just as, for Carnap, external questions concerning which linguistic framework to adopt are not similarly governed by logical rules, but rather require a much less definite appeal to conventional and/or pragmatic considerations, so changes of paradigm in revolutionary science, for Kuhn, do not proceed in accordance with generally agreed upon rules as in normal science, but rather require something more akin to a conversion experience.

It is no wonder, then, as I indicated in Lecture I, that Carnap, in his capacity as editor of the volume of the *Encyclopedia of Unified Science* in which *The Structure of Scientific Revolutions* first appeared, wrote to Kuhn with warm enthusiasm about his then projected work:

> Dear Professor Kuhn:
>
> Thank you very much for sending me your manuscripts. I have read them with great interest, and on their basis I am strongly in favor of your writing a monograph for the Encyclopedia, as you lined out in your letter to Morris of February 13th. I hope that you will find it possible to write your first draft this summer
>
> I believe that the planned monograph will be a valuable contribution to the Encyclopedia. I am myself very much interested in the problems which you intend to deal with, even though my knowledge of the history of science is rather fragmentary. Among many other items I liked your emphasis on the new conceptual frameworks which are proposed in revolutions in science, and, on their basis, the posing of new questions, not only answers to old problems.[50]

And it is no wonder, similarly, that Kuhn, towards the end of his career, regretted the fact that he had earlier interpreted Carnap's letters as expressions of "mere politeness" and fully acknowledged the point that his own philosophical conception is closely akin to the relativized view of the a priori earlier articulated by Reichenbach and Carnap:

---

50. Letter of 12 April 1960; reprinted in Reisch (1991, p. 266).

Though it is a more articulated source of constitutive categories, my structured lexicon [= Kuhn's late version of "paradigm"] resembles Kant's a priori when the latter is taken in its second, relativized sense. Both are constitutive of *possible experience* of the world, but neither dictates what that experience must be. Rather, they are constitutive of the infinite range of possible experiences that might conceivably occur in the actual world to which they give access. Which of these conceivable experiences occurs in that actual world is something that must be learned, both from everyday experience and from the more systematic and refined experience that characterizes scientific practice. They are both stern teachers, firmly resisting the promulgation of beliefs unsuited to the form of life the lexicon permits. What results from respectful attention to them is knowledge of nature, and the criteria that serve to evaluate contributions to that knowledge are, correspondingly, epistemic. The fact that experience within another form of life – another time, place, or culture – might have constituted knowledge differently is irrelevant to its status as knowledge.[51]

All of this very strongly supports my suggestion, I believe, that our best current historiography of science requires us to draw a fundamental distinction between constitutive principles, on the one side, and properly empirical laws formulated against the background of such principles, on the other.

We are now in a position, finally, to connect the main theme of the present Lecture with that of Lecture I. There I argued that Kuhn's theory of scientific revolutions should lead us to resist the idea that philosophy, as a discipline, is to be absorbed into the sciences – either into the natural sciences, as in Helmholtz's psychological conception, or into the mathematical sciences, as in Carnapian *Wissenschaftslogik*. I have now argued that Kuhn's theory of scientific revolutions also requires us to resist the idea, characteristic of Quinean epistemological holism, that there is no fundamental distinction between a priori constitutive principles and properly empirical laws. But Quine's epistemological holism, as we have seen, is itself based on a continued attachment to the idea of a scientific philosophy – to the idea, more specifically, that the scientific epistemology Carnap had attempted with purely logical resources should instead be continued within a naturalized yet equally scientific setting

---

51. Kuhn (1993, pp. 331–2); the remark about "mere politeness" occurs on p. 313.

with the additional resources of behavioristic psychology (compare note 49
above). For Quine, just as for Carnap, the primary ambition is that philosophy should become a science in the strictest sense, in the sense, that is, of the
natural and mathematical sciences. My own suggestion, on the contrary, is
that philosophy need not, and indeed must not, aspire to a scientific status in
this sense. So there is ample room, in particular, for an appeal to the resources
of history as well, and thus for an appeal to the resources of a paradigmatic
Geisteswissenschaft. From this standpoint, the standpoint of our best current
historiography of science, the distinction between constitutive principles and
properly empirical laws (a distinction which slipped through our fingers
within both Quinean epistemology naturalized and, despite Carnap's own
best intentions, within Carnapian *Wissenschaftslogik* as well) again reappears
as most strongly salient indeed.

As I also suggested in Lecture I, my own ambition is to supplement
Kuhnian historiography of science with a parallel, and interrelated appeal to
the concurrent history of scientific philosophy. In order fully to understand
the total ongoing dialectic of our scientific knowledge, I suggest, we need to
replace Kuhn's twofold distinction between normal and revolutionary science
with a threefold distinction between normal science, revolutionary science,
and the philosophical articulation of what we might call meta-paradigms or
meta-frameworks for revolutionary science capable of motivating and sustaining the transition to a new scientific paradigm.[52] In the case of the
Newtonian revolution, for example, the mathematics, mechanics, and physics
together comprising the theory of universal gravitation were also framed, at a
higher level, as it were, by Newton's philosophical encounters with such
thinkers as Descartes and Leibniz over issues concerning the nature of space,
time, matter, force, interaction, and divinity.[53] And in the case of the general

---

52. At the philosophical or meta-scientific level the terminology "meta-framework" is probably
better, so as wholly to avoid the suggestion that a generally agreed upon or common paradigm is
available here. This entire matter is further discussed in the remaining Lectures, and especially in
Part Two, section 4 below.

53. Newton's transformation of the mechanical philosophy (note 46 above) involved, in particular, explicit metaphysical opposition to Descartes on these matters; and metaphysical
disagreements among partisans of the first-level paradigm of the mechanical philosophy (Descartes, Gassendi, Hobbes, Huygens, Spinoza, and Leibniz, for example) also form part of the
necessary "meta-background," as it were, for Newton's radical reconceptualization of that paradigm. Newton's assimilation of the Cambridge Platonism represented by Henry More was
especially important in this regard, for it was here that he acquired a "metaphysics of space"
(including God's relationship to space) sustaining both his commitment to absolute motion and
his use of action-at-a-distance forces mediated by no corporeal influences.

theory of relativity, to take our second example, Einstein's creation of a radically new "geometrized" theory of gravitation was quite explicitly framed within the philosophical debate on the foundations of geometry initiated by Helmholtz and continued by Poincaré.[54] In this sense, just as Kant's conception of constitutive a priori principles has been relativized and generalized within the succeeding tradition of scientific philosophy (and, as I have just observed, within our best current historiography of science as well), the Kantian conception of the peculiarly "transcendental" function of philosophy (as a meta-scientific discipline) must also be relativized and generalized. The enterprise Kant called "transcendental philosophy" – the project of articulating and philosophically contextualizing the most basic constitutive principles defining the fundamental spatio-temporal framework of empirical natural science – must be seen as just as dynamical and historically situated as are the mathematical-physical constitutive principles which are its object.

In place of the Quinean figure of an holistically conceived web of belief, wherein both knowledge traditionally understood as a priori and philosophy as a discipline are supposed to be wholly absorbed into empirical natural science, I would like to suggest an alternative picture of a thoroughly dynamical yet nonetheless stratified or differentiated system of knowledge that can be analyzed, for present purposes, into three main strata or levels. At the base level, as it were, are the concepts and principles of empirical natural science properly so-called: empirical laws of nature, such as the Newtonian law of universal gravitation or Einstein's equations for the gravitational field, which squarely and precisely face the "tribunal of experience" via a rigorous process of empirical testing. At the next or second level are the constitutively a priori principles, basic principles of geometry and mechanics, for example, that define the fundamental spatio-temporal framework within which alone the rigorous formulation and empirical testing of first or base level principles is then possible. These relativized a priori principles constitute what Kuhn calls paradigms: at least relatively stable sets of rules of the game, as it were, that define or make possible the problem solving activities of normal science – including, in particular, the rigorous formulation and testing of properly empirical laws. In periods of deep conceptual revolution it is precisely these constitutively a priori principles which are themselves subject to change – under intense pressure, no doubt, from new empirical findings and especially anomalies. It does not follow, however, that such second level constitutive principles are empirical in the same sense as are the first level principles. On

---

54. See especially Part Two, section 4 below.

the contrary, since here, by hypothesis, a generally agreed upon background framework is necessarily missing, no straightforward process of empirical testing, in periods of deep conceptual revolution, is then possible.[55] And it is precisely here, in fact, that our third level, that of philosophical meta-paradigms or meta-frameworks, plays an indispensable role, by serving as a source for suggestions and guidance – for orientation, as it were – in motivating and sustaining the transition from one paradigm or conceptual framework to another.[56]

None of these three levels are fixed and unrevisable, and the distinctions I am drawing have nothing to do, in particular, with differing degrees of certainty or epistemic security. Indeed, the whole point of the present conception of relativized and dynamical a priori principles is to accommodate the profound conceptual revolutions that have repeatedly shaken our knowledge of nature to its very foundations. It is precisely this revolutionary experience, in fact, that has revealed that our knowledge *has* foundations in the present sense: subject-defining or constitutive paradigms whose revision entails a genuine expansion of our space of intellectual possibilities, to such an extent, in periods of radical conceptual revolution, that a straightforward appeal to empirical evidence is then no longer directly relevant. And it is at this point, moreover, that philosophy plays its own distinctive role, not so much in justifying or securing a new paradigm where empirical evidence cannot yet do so, but rather in clarifying and articulating the new space of intellectual possibilities and making the serious consideration of the new paradigm a reasonable and responsible option. The various strata or levels in our total evolving and interacting system of beliefs are thus not distinguished by differing degrees of epistemic security at all – neither by differing degrees of centrality and entrenchment in the sense of Quine nor by differing degrees of certainty in the more traditional sense – but rather by their radically different functions or roles within the never ending and often entirely unpredictable advance of science as a whole.

---

55. See Part Two, section 2 below.
56. See Part Two, section 4 below.

III

Rationality, Revolution, and the Community of Inquiry

Towards the end of Lecture II I suggested that a dynamical and relativized version of the Kantian conception of a priori constitutive principles (originating within the logical empiricist tradition), as organizing and making possible properly empirical knowledge of nature, is very strongly supported by what I called "our best current historiography of science" – namely, Thomas Kuhn's theory of the nature and character of scientific revolutions. Kuhn's notion of a "paradigm" (later replaced by what he calls a "structured lexicon") governing a particular stage or episode of normal science functions precisely as a language or conceptual framework within which the problem solving activities of this kind of science then proceed. In Rudolf Carnap's terminology, normal science is occupied exclusively with *internal* questions, which are answerable, at least in principle, by means of what Carnap calls the logical rules definitive of the framework in question. In periods of revolutionary science, by contrast, it is precisely such a conceptual framework that is now in question, so that we have no generally agreed upon system of logical rules capable, at least at the time, of settling the issue. A transition to a new paradigm or conceptual framework thus definitely exceeds the bounds of normal science: in Carnap's terms, it involves us with an *external* question concerning which form of language or system of logical rules we should now adopt. Until the new conceptual framework itself becomes well established, therefore, we must proceed entirely without the benefit of the generally agreed upon and taken for granted rules definitive of normal science.

This close parallel between what I am calling the relativized, yet still constitutive a priori, on the one side, and Kuhn's theory of scientific revolutions, on the other, implies, however, that the former gives rise to the same problems and questions concerning the ultimate rationality of the scientific enterprise that are all too familiar in the post-Kuhnian literature in history, sociology, and philosophy of science. For, on the one hand, since there appear

to be no generally agreed upon logical rules governing the transition to a revolutionary new scientific paradigm or conceptual framework, there would
seem to be no sense left in which such a transition can still be viewed as rational, as based on good reasons. Non-rational factors, having more to do with
persuasion or conversion than rational argument, must necessarily be called
in to explain the transition in question. And, on the other hand, since only
such a non-rational commitment of the scientific community as a whole can
explain the acceptance of a particular scientific paradigm at a given time, it
would appear that the only notion of scientific rationality we have left is a relativized, sociological one according to which all there ultimately is to
scientific rationality – and thus to scientific knowledge – is the otherwise
arbitrary commitment of some particular social community or group to one
particular paradigm or framework rather than another.

The underlying source of this post-Kuhnian predicament, as we have seen,
is the breakdown of the original Kantian conception of the a priori. Kant
takes the fundamental constitutive principles framing Newtonian mathematical science as expressing timelessly fixed categories and forms of the human
mind. Such categories and forms, for Kant, are definitive of human rationality as such, and thus of an absolutely *universal* rationality governing all
human knowledge at all times and places. Moreover, as we have also seen, this
conception of an absolutely universal human rationality realized in the fundamental constitutive principles of Newtonian science made perfectly good
sense in Kant's own time, when the Newtonian conceptual framework was the
only paradigm for what we now call mathematical physics the world had yet
seen. Now that we have irretrievably lost this position of innocence, however,
it would appear that the very notion of a truly universal human rationality
must also be given up. It would appear that there is now no escape from the
currently fashionable slogan "all knowledge is local."

Kuhn himself raises the threat of sociological conceptual relativism in the
very passage (which we looked at towards the end of Lecture II) where he
makes the parallel between his conception of science and the relativized version of the Kantian a priori fully explicit. Thus, in the passage to which note 51
above is appended, Kuhn observes that his "structured lexicon resembles
Kant's a priori ... in its second, relativized sense" in being "constitutive of
possible experience of the world." And he concludes by remarking that "[t]he
fact that experience within another form of life – another time, place, or
culture – might have constituted knowledge differently is irrelevant to its status as knowledge." In this last sentence, then, Kuhn simultaneously raises the
threat of conceptual relativism and shows, nonetheless, that he wants to mini-

mize the relativistic implications of his view. I will come back to Kuhn's attempt to distance himself from relativism in a moment, but I first want to remind you of how enthusiastically some of his followers have willingly embraced such implications.

Perhaps the most explicit and provocative defense of sociological conceptual relativism is found in the "sociology of scientific knowledge" of the so-called Edinburgh School, both the theoretical work of the founders of this school, Barry Barnes and David Bloor, and the applied work of social historians of science self-consciously working in the Edinburgh tradition, such as Steven Shapin and Simon Schaffer.[57] For the underlying philosophical theme or agenda framing all of this work is the idea that the traditional notions of rationality, objectivity, and truth reduce, in the end, to local socio-cultural norms conventionally adopted and enforced by particular socio-cultural groups. In a book devoted primarily to a discussion of the sociological implications of Kuhnian historiography, Barnes expresses this idea as follows:

> Science is not a set of universal standards, sustaining true descriptions and valid inferences in different specific cultural contexts; authority and control in science do not operate simply to guarantee an unimpeded interaction between 'reason' and experience. Scientific standards themselves are part of a specific form of culture; authority and control are essential to maintain a sense of the reasonableness of that specific form. Thus ... science should be amenable to sociological study in fundamentally the same way as any other form of knowledge or culture.[58]

And, in a well-known jointly authored paper, entitled "Relativism, Rationalism and the Sociology of Knowledge," Barnes and Bloor explain that

> [The relativist] accepts that none of the justifications of his preferences can be formulated in absolute or context-independent terms. In the last analysis, he acknowledges that his justifications will stop at some principle or alleged matter of fact that has only local credibility. ... For the relativist there is no sense attached to the idea that some standards or beliefs are really rational as distinct from merely locally accepted as such. Because he thinks

---

57. Here I of course have in mind especially Shapin and Schaffer (1985)
58. Barnes (1982, p. 10).

that there are no context-free or super-cultural norms of rationality he does not see rationally and irrationally held beliefs as
making up two distinct and qualitatively different classes of
things.[59]

Thus the two fundamental principles of the Edinburgh program for a sociology of scientific knowledge – that scientific knowledge as such is amenable to
sociological study and that both "rational" and "irrational" beliefs should be
explained "symmetrically" – are directly based on a version of philosophical
conceptual relativism derived from Kuhn. They are based, in the end, on the
idea that there are no "universal standards" of human reason, no "context-free
or super-cultural norms of rationality."[60]

Yet Kuhn himself, as just noted, rejects such relativistic implications of his
views. He continues to hold, in a self-consciously traditional vein, that the
evolution of science is a rational and progressive process despite the revolutionary transitions between scientific paradigms which are, as he also claims,
absolutely necessary to this process. The scientific enterprise, Kuhn suggests,
is essentially an instrument for solving a particular sort of problem or
"puzzle" – for maximizing the quantitative match between theoretical predictions and phenomenological results of measurement. Given this, however,
there are obvious criteria or "values" – such as accuracy, precision, scope,
simplicity, and so on – that are definitive of the scientific enterprise as such.
Such values are constant or permanent across scientific revolutions or paradigm-shifts, and this is all we need to secure the (non-paradigm-relative)
rationality of scientific progress:

> [W]hether or not individual practitioners are aware of it, they
> are trained to and rewarded for solving intricate puzzles – be
> they instrumental, theoretical, logical, or mathematical – at the
> interface between their phenomenal world and their commu-
> nity's beliefs about it. ... If that is the case, however, the
> rationality of the standard list of criteria for evaluating scientific
> belief is obvious. Accuracy, precision, scope, simplicity, fruitful-
> ness, consistency, and so on, simply *are* the criteria which puzzle
> solvers must weigh in deciding whether or not a given puzzle

59. Barnes and Bloor (1982, pp. 27–8).
60. For an extended discussion of the sociology of scientific knowledge and its historical-philo-
sophical context see Friedman (1998).

about the match between phenomena and belief has been solved. ... To select a law or theory which exemplified them less fully than an existing competitor would be self-defeating, and self-defeating action is the surest index of irrationality. ... As the developmental process continues, the examples from which practitioners learn to recognize accuracy, scope, simplicity, and so on, change both within and between fields. But the criteria that these examples illustrate are themselves necessarily permanent, for abandoning them would be abandoning science together with the knowledge which scientific development brings. ... Puzzle-solving is one of the families of practices that has arisen during that evolution [of human practices], and what it produces is knowledge of nature. Those who proclaim that no interest-driven practice can properly be identified with the rational pursuit of knowledge make a profound and consequential mistake.[61]

Thus, although the process of scientific development is governed by no single conceptual framework fixed once and for all, science, at *every* stage, still aims at a uniform type of puzzle-solving success, Kuhn suggests, relative to which *all* stages in this process (including transitions between conceptual frameworks) may be judged. And there is then no doubt at all, Kuhn further suggests, that science, throughout its development, has become an increasingly efficient instrument for achieving this end. In this sense, therefore, there is also no doubt at all that science as a whole is a rational enterprise.

Unfortunately, Kuhn's attempt to defend the rationality of scientific knowledge from the threat of conceptual relativism cannot, I believe, be judged a success. In the first place, there are powerful reasons arising from Kuhn's own historiography for doubting whether any such puzzle-solving criteria are really permanent across revolutionary scientific change. Whereas puzzle-solving accuracy and precision, for example, have always been criteria or "virtues" of astronomical practice, this has emphatically not been the case in other scientific fields. Indeed, one of the main motivations of the scientific revolution of the sixteenth and seventeenth centuries was precisely to make such accuracy and precision criteria for successful *terrestrial* mechanics, just as they had always been so for *celestial* mechanics. (In Aristotelian-Scholastic natural philosophy, by contrast, quantitative accuracy and precision do not

---

61. Kuhn (1993, pp. 338–9).

play a central role in the terrestrial realm.) Moreover, although quantitative accuracy and precision (in both the celestial and the terrestrial realm) thus became values or ends of the scientific enterprise in the sixteenth and seventeenth centuries, they were not really able to function effectively as such until the transition between the eighteenth and nineteenth centuries that has been dubbed the "second scientific revolution."[62] For it was not until this time, as a matter of fact, that anything like quantitative accuracy of predictions in such fields as chemistry, electricity, magnetism, heat, and so on could actually be achieved. Here one suspects that Kuhn is using a rather anachronistic characterization of the term "science" (one arising towards the latter part of the nineteenth century) that is itself by no means constant or permanent accross the entire developmental process his historiography aims to describe.

In the second place, even if we admit that there are constant or permanent criteria or values definitive of scientific success, it remains entirely obscure how there can be an "uncommitted" or paradigm-independent standpoint for rationally assessing the satisfaction of such criteria. Different paradigms, according to Kuhn, represent different conceptual frameworks or languages. Practitioners of one paradigm use a framework of concepts and principles incommensurable or non-intertranslatable with that of another paradigm, and it is only relative to one paradigm or another that the practitioners in question can coherently describe and experience their respective worlds. How can the practitioners of one framework even *understand* the claim, therefore, that another framework better satisfies the criteria or values of scientific success? Consider the choice between special relativity and the classical aether based theory developed by Lorentz, for example. Both theories are quantitatively accurate to the same degree – indeed, as we have said, there is an important sense in which they are both empirically and mathematically equivalent. According to the Einsteinians, however, special relativity better satisfies the virtue of simplicity, since it dispenses with the aether entirely. But, from the point of view of the aether theorists, Einstein's theory thereby dispenses as well with the one spatio-temporal-mechanical framework – that of classical, Newtonian mechanics – within which alone it is possible to think coherently about spatio-temporal-mechanical phenomena at all. Einstein's theory, from this point of view, is simply not a coherent possibility. So there can be no

---

62. See Bellone (1976/80). Kuhn (1976/77) touches on important aspects of this transition himself. I return to it briefly in Part Two, section 5 below, especially in the paragraph to which notes 65 and 66 are appended.

room for judgements of (mere) simplicity here until *after* we have made the conceptual transition.[63]

Finally, and in the third place, even if we accept Kuhn's puzzle-solving account of scientific rationality as unproblematic, there is considerable room for doubt, I believe, whether it can possibly do justice to the issue of conceptual relativism arising in the wake of Kuhn's own historiographical work. For it is surely uncontroversial that the scientific enterprise as a whole has in fact become an ever more efficient instrument for puzzle-solving in this sense – for maximizing quantitative accuracy, precision, simplicity, and so on in adjusting theoretical predictions to phenomenological results of measurement. What is controversial, rather, is the further idea that the scientific enterprise thereby counts as a privileged model or exemplar of rational knowledge of – rational inquiry into – nature. That the scientific enterprise thus represents a model or exemplar of human rationality as such was of course a main pillar of the philosophy of the Enlightenment, and it was precisely this Enlightenment vision, as we have seen, that Kant made fully explicit through the idea that the fundamental constitutive principles of Newtonian mathematical physics encapsulate absolutely fixed standards of a permanent and universal human rationality. Yet the succeeding scientific revolutions of the nineteenth and twentieth centuries have shown that the Newtonian framework is by no means fixed and universal in this sense; and it is precisely this phenomenon, moreover, that has led to our current predicament vis-à-vis conceptual relativism. Pointing to the obvious (if not entirely unproblematic) fact that science has nonetheless continued to increase its quantitative accuracy, precision, and so on appears to be a quite inadequate response to the full force of this relativistic challenge.[64]

The inadequacy of Kuhn's response rests, in the end, on a failure clearly to distinguish between two very different aspects of human rationality. Following terminology introduced by Jürgen Habermas, I will call the first, *instrumental rationality,* and the second, *communicative rationality:*

---

63. It is worth noting that Kuhn (1973/77, pp. 334–9) touches on both this point and that of the preceding paragraph.

64. As I explain in Part Two, sections 2 and 3 below, the problems raised in this and the preceding paragraph are really two aspects of a single problem for Kuhn. The crucial point is that we have to distinguish between conceiving scientific theories as mere predictive "black boxes" and conceiving them as genuinely meaningful systems of propositions sustaining a system of evidential relationships. Kuhn's notion of paradigm (like Carnap's notion of linguistic framework) involves a relativized version of the latter conception, and it thus raises a threat of conceptual relativism that is in no way alleviated by appealing to mere "black box" predictive success.

If we proceed from the non-communicative employment of propositional knowledge in goal-directed actions, we find a prior decision in favor of that concept of *cognitive-instrumental rationality*, which, in empiricism and beyond, has strongly influenced the self-understanding of modernity. It carries the connotations of successful self-maintenance, made possible by informed mastery of, and intelligent adaptation to, conditions of a contingent environment. By contrast, if we proceed from the communicative employment of propositional knowledge in linguistic actions, we find a prior decision in favor of a wider concept of rationality tied to the older representation of logos. This concept of *communicative rationality* carries connotations that ultimately trace back to the central experience of the non-coercively uniting, consensus creating power of argumentative speech, in which different participants overcome their initially subjective points of view, and, thanks to the commonality of reasonably motivated convictions, assure themselves simultaneously of the unity of the objective world and the intersubjectivity of their context of life.[65]

Instrumental rationality thus refers to our capacity to engage in effective means-ends deliberation or reasoning aimed at maximizing our chances of success in pursuing an already set end or goal. It takes the goal in question as given, and it then attempts to adjust itself to environmental circumstances in bringing this desired state of affairs into existence in the most efficient way possible. This is the sense of rationality David Hume had in mind, of course, in his famous dictum that "reason is, and ought only to be the slave of the passions." Communicative rationality, by contrast, refers to our capacity to engage in argumentative deliberation or reasoning with one another aimed at bringing about an agreement or consensus of opinion. It starts with the idea that disagreement about some particular matter of mutual interest is problematic or undesirable, and that resolution of such disagreement must appeal to patterns of argument or reasoning acceptable to all parties in the dispute.

---

65. Habermas (1981/84, vol. 1, chapter 1, section 1.A). This distinction is parallel to that between the "rational" and the "reasonable" emphasized in Rawls (1993/96, Lecture II, § 1), where it is associated with the Kantian distinction between hypothetical and categorical imperatives. The second (1996) edition of Rawls's book contains his side of an important debate with Habermas concerning the foundations of democratic theory. I hope to discuss the implications of the present conception of scientific rationality for this debate in future work.

Then, on the basis of precisely such mutually recognized principles of reasoning, it attempts to reach consensus, or at least mutual understanding, yielding a resolution of the dispute acceptable to all. The Enlightenment faith in human rationality, as expressed most explicitly by Kant, for example, is rooted in the idea that this kind of rationality – communicative rationality – provides access to absolutely universal principles of reasoning common to all human beings as such.

Instrumental rationality is in an important sense private or subjective. It takes one or another end or goal as given, and it then inquires into the best means for achieving this goal. Human ends, however, are notoriously diverse and variable, and so there can be no ground for a truly universal rationality within purely instrumental reason. This, for example, is why Kuhn's attempt to find permanent criteria or values held constant throughout the development of science necessarily fails. Quantitative accuracy and precision were not widespread ends of inquiry within Aristotelian-Scholastic natural philosophy; and, even though they became such ends with the scientific revolution of the sixteenth and seventeenth centuries, they did not become meaningful widespread ends of inquiry – that is, practically achievable such ends – until the "second scientific revolution" of the late eighteenth and early nineteenth centuries; simplicity, I would argue, only became a central and effective end of inquiry with the relativistic revolution of the early twentieth century; and so on.[66]

Communicative rationality, by contrast, is essentially public or intersubjective. It aims, by its very nature, at an agreement or consensus based on mutually acceptable principles of argument or reasoning shared by all parties in a dispute. This is the kind of rationality underwritten by a given scientific paradigm or conceptual framework, whose function is precisely to secure an agreement on fundamental constitutive principles – in Kuhn's words, a "firm research consensus" that is "universally received" within a particular scientific community. Normal science, for Kuhn, is then entirely based on this kind of "firm research consensus."[67] But it is precisely because there is also revolu-

---

66. This way of putting the matter can be seriously misleading. There is no reason, for example, that the end of purely instrumental "black box" predictive success could not be adopted as a universally shared goal of scientific research (indeed, it appears that this is in fact the case). The point, rather, is that communicative rationality – as opposed to merely instrumental "black box" predictive success – places us within a shared public language or linguistic framework, and thus within a shared "space of reasons," where theoretical propositions can be meaningfully subject to evidential evaluation: see again Part Two, section 2 below.

67. See note 21 above.

tionary science – radical transitions between incommensurable or non-intertranslatable conceptual frameworks – that the idea of a truly universal rationality (that is, a truly universal *communicative* rationality) is now threatened.

In contrast to Kuhn's failure clearly to distinguish between these two quite different concepts of rationality, Carnap's philosophy of formal languages or linguistic frameworks centrally involves a very sharp distinction between them. Internal questions, for Carnap, are answerable, at least in principle, by reference to the logical rules of one or another given framework. Here we have genuine theoretical questions where notions of "correct" and "incorrect," "true" and "false" clearly and unproblematically apply. Investigators who share a given linguistic framework can then engage in genuinely cognitive or theoretical disputes about precisely such internal questions. But external questions (questions essentially involving a choice between linguistic frameworks) are not genuinely rational in this sense – that is, in Habermas's terminology, they are not similarly *communicatively* rational. Here, for Carnap, we are faced with merely pragmatic or instrumental questions about the suitability or appropriateness of one or another framework for some or another given purpose. This means, first, that the answers to external questions cannot be judged by yes/no dichotomies such as "correct" and "incorrect," "true" and "false," but rather always involve matters of degree; and it also means, second, that the answers to external questions are necessarily relative to the given ends or goals of particular investigators (especially cautious investigators fearful of the possibility of contradiction, for example, may prefer the weaker rules of intuitionistic logic, whereas those more interested in ease of physical application may prefer the stronger rules of classical logic).[68] Unfortunately, however, although Carnap thus clearly separates instrumental from communicative rationality, he does not address the threat of conceptual relativism afflicting the latter concept of rationality that directly flows from his own embrace of the relativized conception of constitutive a priori principles – principles defining what "correct" and "incorrect," "true" and "false" mean *relative to* one or another linguistic framework. Carnap's relativism is explicitly addressed, rather, to the dissolution of what he takes to be

---

68. Again, the first point is the more important one, for there is no reason a given community of investigators cannot agree on a given goal (and, conversely, there is also no guarantee, of course, that all investigators will share a common linguistic framework). The crucial point is that external questions are not subject to evaluation as true or false, correct or incorrect, and are thus not subject to rational evidential evaluation according to the internal standards of a (single) linguistic framework.

fruitless disputes within philosophy as a discipline, and he nowhere considers seriously the relativistic predicament arising in the wake of Kuhn's work on scientific revolutions.[69] If pressed he would probably (and reasonably) respond that the sciences can here, as elsewhere, take care of themselves.

Yet we post-Kuhnians, I fear, cannot rest content with such a response, especially when an explicitly relativistic philosophical agenda now informs a significant segment of research in the history, sociology, and philosophy of science. It is true that the sciences (at least the natural sciences) can take care of themselves (although even some natural scientists, as I observed in Lecture I, now feel threatened by the widespread relativistic tide). But we in philosophy (and in the humanities more generally) are nonetheless faced with a radically new question as to how this is possible. We are faced with radically transformed versions, that is, of Kant's original questions "How is pure mathematics possible?" and "How is pure natural science possible?" – for our problem is clearly and responsibly to address such questions in an intellectual environment where Kant's original conception of an absolutely fixed and universal communicative rationality constitutively framing the properly empirical advance of natural science has unequivocally broken down once and for all.

•

Let us begin by reminding ourselves of the characteristic feature distinguishing the scientific enterprise from other areas of intellectual and cultural life – a feature that figures centrally, as we have just seen once again, in Kuhn's theory of scientific revolutions. In the scientific enterprise, unlike other areas of intellectual and cultural life, there is such a thing as normal science – periods of "firm research consensus" in which a given paradigm, conceptual framework, or set of rules of the game, as it were, is "universally received" by all practitioners of a given field or discipline. Indeed, as Kuhn points out, science as such comes into being when a pre-paradigm state of diverse and competing schools of thought is replaced by the single paradigm or framework characteristic of normal science. In the sciences, then, we are actually able to achieve a situation of communicative rationality far exceeding that possible in other areas of intellectual and cultural life; and this is undoubtedly the rea-

---

69. See note 18 above, together with the paragraph to which it is appended. It is true that Carnap (1963, p. 921), for example, explicitly associates his philosophical conception with the idea of a scientific revolution, but this is never the primary application he has in mind.

son that the scientific enterprise has been taken to be a particularly good model or exemplar of human rationality from the Enlightenment on (if not much earlier).[70]

The communicative rationality characteristic of normal science is of course limited to what we might call intra-paradigm or intra-framework rationality, and it does not, by itself, address the threat of conceptual relativism. On the contrary, it is precisely because this kind of rationality is defined only relative to one or another paradigm or framework that the threat of conceptual relativism then arises. However, it is not only the case, as Kuhn rightly emphasizes, that the scientific enterprise differs from other areas of intellectual and cultural life within particular stages or episodes of normal science, it also differs fundamentally in the way in which it treats transitions between such stages – in the way, that is, in which it handles revolutionary transitions. Just as the scientific enterprise aims for, and successfully achieves, agreement or consensus *within* particular paradigms, there is also an important sense in which it aims for, and successfully achieves, agreement or consensus *across* different paradigms. In the sciences, unlike other areas of intellectual and cultural life, we are never in a position simply to throw out all that has gone before (if only rhetorically) and start again anew with an entirely clean slate.[71] On the contrary, even, and indeed especially, in periods of deep conceptual revolution, we still strive to preserve what has gone before (the preceding paradigm or framework) as far as possible.

This process is most clearly evident in the more advanced mathematical sciences, where the effort at preservation takes the precise form of exhibiting the preceding paradigm or framework as an approximate special case – valid in precisely defined special conditions – of the succeeding paradigm. Special relativistic mechanics approaches classical mechanics in the limit as the velocity of light goes to infinity (or, equivalently, as we consider only velocities very small in comparison with that of light); variably curved Riemannian geome-

---

70. At Stanford Richard Rorty asked whether other cultural enterprises, such as jurisprudence, for example, might not have achieved an equal situation of communicative rationality. As far as I can judge, however, although this may be true within particular national traditions during particular historical periods, I do not see how anything approaching the trans-national and trans-historical communicative rationality characteristic of the mathematical exact sciences has been attained anywhere else. Indeed, within the Western intellectual and cultural tradition this situation appears to extend back to the ancient Greeks. (But it is worth noting, once again, that other intellectual and cultural enterprises may exhibit widespread agreement on "paradigms" in the sense of exemplary works or achievements: see note 23 above.)

71. At least rhetorically, for example, this happens not infrequently in the discipline of philosophy.

try approaches flat Euclidean geometry as the regions under consideration become infinitely small; Einstein's general relativistic field equations of gravitation approach the Newtonian equations for gravitation as, once again, the velocity of light goes to infinity.[72] Indeed, even in the transition from Aristotelian terrestrial and celestial mechanics to classical terrestrial and celestial mechanics we find a similar, if not quite so precise, relationship. From an observer fixed on the surface of the earth we can construct a system of lines of sight directed towards the heavenly bodies; this system is spherical, isomorphic to the celestial sphere of ancient astronomy, and the motions of the heavenly bodies therein are indeed described, to a very good approximation, by the geocentric system favored by Aristotle (that is, the Eudoxian system of homocentric spheres). Moreover, in the sublunary region close to the surface of the earth, where the earth is by far the principal gravitating body, heavy bodies do follow straight paths directed towards the center of the earth, again to an extremely good approximation. In all three revolutionary transitions (Aristotelian natural philosophy to classical mechanics, classical mechanics to special relativity, and special relativity to general relativity), therefore, key elements of the preceding paradigm are preserved, as far as possible, in the succeeding paradigm, so that, in particular, we are able to depict ourselves as extensively agreeing with the practitioners of the preceding paradigm (again as far as possible) at the very same time that we are disagreeing fundamentally on the concepts and principles of the paradigm itself.

Of course this kind of relationship of successive approximation between paradigms, and the resulting notion of *inter*-paradigm agreement, is highly anachronistic or "Whiggish," in that it is constructed wholly from the point of view of the successor paradigm and, in truly revolutionary cases, uses concepts and principles that simply do not exist from the point of view of the preceding paradigm. The four dimensional Minkowski space-time of special relativity that approaches what we now call Newtonian space-time (note 43 above) as the velocity of light goes to infinity does not even exist as a conceptual possibility within the essentially three dimensional framework of classical mechanics; the concept of a variably curved manifold does not yet exist within the essentially Euclidean framework of all pre-nineteenth century mathematics; and, within Aristotelian celestial and terrestrial mechanics, the

---

72. As indicated in note 20 above, Kuhn (1962/70, pp. 101–2) explicitly denies that classical mechanics can be logically derived from relativistic mechanics in the limit of small velocities, on the grounds, primarily, that "the physical referents" of the terms of the two theories are different. Here, however, I am merely pointing to a purely mathematical fact about the corresponding mathematical structures.

key conception, definitive of mature classical mechanics, of an infinite Euclidean universe wherein natural inertial motion continues indefinitely in a Euclidean straight line is also quite impossible. Practitioners of the earlier paradigm are thus not even in a position *to understand* the agreement claimed by practitioners of the later paradigm, and this is why, from an historical point of view, such retrospective reconstructions are quite inadequate for properly understanding the earlier conceptual framework itself. Indeed, if we acknowledge the point that the meanings of the terms of any paradigm or framework are fundamentally shaped by the constitutive principles of the framework in question, it follows that there is an important sense in which practitioners of the later framework, in thus reconstructing the previous framework, also thereby fail to understand it. What is exhibited as an approximate special case, being constructed entirely within the new conceptual framework, is a rational reconstruction of the earlier framework and not the earlier framework itself.[73]

Nevertheless, from a properly historical, non-anachronistic and non-Whiggish point of view, we can then take a further step. We can exhibit the historical evolution by which the new concepts and principles gradually emerge through successive transformations of the old concepts and principles. Practitioners of succeeding paradigms are not helpfully viewed as members of radically disconnected speech communities, as speakers of radically diverse languages, as it were, such that only intensive immersion in an entirely foreign culture can possibly take us from one paradigm to another. On the contrary, successive paradigms emerge precisely from one another, as succeeding stages in a common tradition of cultural change. In this sense, they are better viewed as different evolutionary stages of a single language rather than as entirely separate and disconnected languages.[74]

Consider, for example, the transition from Aristotelian terrestrial and celestial mechanics to classical mechanics. The Aristotelian conceptual framework is based on, first, Euclidean geometry (which of course serves as a primary exemplar of rational knowledge for everyone until the early nineteenth

---

73. This captures the sense in which successive frameworks are "incommensurable" in the Kuhnian sense: see Part Two, section 3 below, especially note 37 and the paragraph to which it is appended. That the convergence in question yields only a purely retrospective *reinterpretation* of the original theory is a second (and related) point Kuhn makes in the discussion cited in note 72 above, where he points out (1962/70, p. 101) that the laws derived as special cases in the limit within relativity theory "are not [Newton's] unless those laws are reinterpreted in a way that would have been impossible until after Einstein's work." (I am indebted here and in note 72 to comments from Michael Heidelberger.)

74. See again Part Two, section 3 below, especially note 33.

century), second, a background conception of a hierarchically and teleologically organized universe, third, conceptions of natural place and natural motion appropriate to this universe. Thus, in the terrestrial realm, heavy bodies naturally move in straight lines towards their natural place at the center of the universe, and, in the celestial realm, the heavenly bodies naturally move uniformly in circles around this center. The conceptual framework of classical physics then retains Euclidean geometry, but eliminates the hierarchically and teleologically organized universe together with the accompanying conception of natural place. We thereby obtain an infinite, homogeneous and isotropic universe in which all bodies naturally move uniformly along straight lines to infinity. But how did we arrive at this conception? An essential intermediate stage is Galileo's celebrated treatment of free fall and projectile motion. For, although Galileo indeed discards the hierarchically and teleologically organized Aristotelian universe, he retains – or better, transforms – key elements of the Aristotelian conception of natural motion. Galileo's analysis is based on two concepts of natural motion: what he calls naturally accelerated motion directed towards the center of the earth, and what he calls uniform or equable motion directed at right angles to the former motion. Unlike our modern concept of *rectilinear* inertial motion, however, this Galilean counterpart is uniformly *circular* – traversing points equidistant from the center at constant speed. Yet, in relatively small regions near the earth's surface, this uniform circular motion is quite indistinguishable from uniform rectilinear motion, and this is how Galileo can treat it mathematically as rectilinear to an extremely good approximation. And it is in precisely this way, therefore, that the modern conception of rectilinear natural inertial motion is actually continuous with the preceding Aristotelian conception of natural motion.[75]

Consider, as a second example, the transition from classical physics to Einsteinian relativity theory. This example is considerably more complex, and a proper treatment would have to attend, in addition, to the so-called "second scientific revolution" bridging the transition from the eighteenth to the nineteenth century (note 62 above). I can perhaps say enough here, however, to indicate some of the essential steps. Relativity theory, as I have suggested more than once, emerges not only from the late nineteenth and early twentieth century work on the electrodynamics of the aether associated with the names of Lorentz, Fitzgerald, and Poincaré, but also from revolutionary work throughout the nineteenth century on the foundations of geometry. The former yielded, well before Einstein, what we now know as the Lorentz group,

---

75. Here I am indebted to discussion with William Newman.

which gives the mathematical form of the transformations between frames of reference moving relatively to one another and to the aether. These transformations comprise the mathematical core of special relativity as well, and this is the sense, as I have also suggested, in which special relativity is both empirically and mathematically equivalent to the preceding aether theory. But Einstein's revolutionary move lies in his *interpretation* of these transformations – as not simply representing special dynamical properties of electromagnetically constructed objects as they move relative to the aether, but rather as constitutive of the fundamental geometrical-kinematical framework of what we now call Minkowski space-time. For Einstein, we might say, the Lorentz transformations change from being properly empirical laws (that is, dynamical laws) governing a particular variety of force (electro-magnetic force) to geometrical-kinematical constitutive principles articulating a radically new type of space-time structure—a space-time structure which, from a classical point of view, is simply incoherent.

How was it possible for Einstein to make this revolutionary move? Here nineteenth century work on the foundations of geometry, especially that of Poincaré, was absolutely crucial. We know that Einstein was intensively reading Poincaré's *Science and Hypothesis* immediately before his revolutionary breakthroughs in 1905 (note 27 above). In this work, as is well known, Poincaré not only deals incisively with the current situation in electrodynamics, he also formulates an entirely original perspective on the nineteenth century development of non-Euclidean geometries (which essentially includes Poincaré's own deep mathematical work on this subject). Poincaré rejects the view, defended by Helmholtz, that the choice between Euclidean and non-Euclidean geometries is empirical, for he continues to maintain, following Kant, that geometry functions rather as a constitutive framework making properly empirical discoveries first possible. We now know, however, that there is more than one such constitutive framework, and so the choice between them, Poincaré argues, is based on a convention – it rests, in the end, on the greater mathematical simplicity of the Euclidean system (and not on some innate necessity of the human mind). Einstein, we might say, then applies this very same point of view to the situation in electrodynamics. Just as there is no longer a single, uniquely privileged geometrical framework (built in to the structure of the human mind), there is also no single, uniquely privileged geometrical-kinematical framework (describing, as Minkowski puts it, the geometry of space-time). In particular, we are now free to adopt a new convention – again based on mathematical simplicity – that replaces the classical constitutive framework with the special relativistic one. In this way,

Einstein's introduction of a radically new constitutive framework for space, time, and motion again grew naturally out of, and is thus quite continuous with, the preceding framework it replaced. Moreover, although I cannot discuss the transition to general relativity here, this step also grows naturally out of the nineteenth century tradition in the foundations of geometry, when read in the context of the new mechanics of the special theory: the key move to a non-Euclidean geometry of variable curvature in fact results from applying the Lorentz contraction to the geometry of a rotating disk, as Einstein simultaneously delicately positions himself within the geometrical debate between Helmholtz and Poincaré.[76]

So far we have made two consequential observations about *inter*-paradigm relations, at least in so far as the most abstract and general constitutive principles of geometry and mechanics are concerned. First, from a point of view internal to the development of these sciences, we see that earlier constitutive frameworks are exhibited as limiting cases, holding approximately in certain precisely defined special conditions, of later ones. Second, from a properly historical point of view, we see also that, although there is indeed an important sense in which succeeding paradigms are incommensurable or non-intertranslatable, the concepts and principles of later paradigms still evolve continuously, by a series of natural transformations, from those of earlier ones. Now, putting these two observations together, we are in a position to add, from a philosophical point of view, that we can thus view the evolution of succeeding paradigms or frameworks as a convergent series, as it were, in which we successively refine our constitutive principles in the direction of ever greater generality and adequacy. When we move from the Aristotelian framework to that of classical physics, we retain Euclidean geometry intact, discard the hierarchically and teleologically organized spherical universe, and modify the Aristotelian conception of natural motion – in such a way that we retain the idea, in particular, that there is a fundamental state of natural motion following privileged paths of the underlying geometry. When we move from classical physics to special relativity, we again retain Euclidean geometry, and also, of course, the law of inertia, but we now move from a three-plus-one dimensional to an essentially four dimensional spatio-temporal structure – which, however, yields the older three-plus-one dimensional structure as a limiting case. Finally, when we move to general relativity, we replace Euclidean geometry with the more general structure of *infinitesimally*

---

76. Both of these transitions – to special and to general relativity – are discussed in considerably more detail in Part Two, sections 3 and 4.

Euclidean geometry (that is, Riemannian geometry), and, at the same time, we modify the fundamental state of natural motion, in accordance with the principle of equivalence, so that all motions governed solely by gravitation now count as natural – as straightest possible paths or geodesics of the underlying four dimensional space-time geometry.

Kant, in the eighteenth century, viewed the future evolution of science as not only framed by fixed and unrevisable constitutive principles (in particular, the principles of Newtonian mathematical physics), but also as directed by what he called *regulative* principles – of simplicity, unity, and so on – which guide scientific progress without constitutively constraining it. In this way science approaches an ideal state of completion, but only as a *focus imaginarius*, or regulative ideal of reason, for which there is no guarantee at all that it will ever be actually achieved.[77] Indeed, we can never take our actual science to have reached this ideal, for this would halt scientific progress dead in its tracks; and, at the same time, we are always under the obligation to keep seeking it, for failing to do so would again make science as we know it impossible. What I am now suggesting is that we should adapt Kant's conception of the regulative use of reason to what he called the constitutive domain as well – so as to attain, in particular, an inter-paradigm, trans-historical universality within *this* domain. Since deep conceptual revolutions or paradigm-shifts are a fact of scientific life (and, I would argue, a necessity), we are never in a position to take our present constitutive principles as truly universal principles of human reason – as fixed once and for all throughout the evolution of science. We can imagine, however, that our present constitutive principles represent one stage of a convergent process, as it were, in that they can be viewed as approximations to more general and adequate constitutive principles that will only be articulated at a later stage. We can thus view our present scientific community, which has achieved temporary consensus based on the communicative rationality erected on its present constitutive principles, as an approximation to a final, ideal community of inquiry (to use an obviously Peircean figure) that has achieved a universal, trans-historical communicative rationality on the basis of the fully general and adequate constitutive principles reached in the ideal limit of scientific progress.[78] Indeed, we *must* view

---

77. The Kantian conception of regulative principles (and of the regulative use of reason) is first developed in the Appendix to the Transcendental Dialectic in the *Critique of Pure Reason*; it is further developed in the Introductions to the *Critique of Judgement*.

78. Peirce explained his conception of scientific truth – defined as what the scientific community agrees to in the ideal limit – in two articles first published in the *Popular Scientific Monthly* in 1877 and 1878: see Hartshorne and Weiss (1931–35, vol. 5, §§ 358–87, 388–410).

our present scientific community as an approximation to such an ideal community, I suggest, for only so can the required inter-paradigm notion of communicative rationality be sustained.

This conception of a convergent sequence of constitutive principles, governed by what Kant called the regulative use of reason, is prefigured in the thought of the Marburg School of neo-Kantianism of the late nineteenth and early twentieth centuries, and, in particular, in the work of Ernst Cassirer, the last important representative of this tradition. Thus, after giving the example of how even the fundamental principles of Newtonian mechanics "may not hold as simply unalterable dogmas," but may be transformed into radically new principles, Cassirer points out that "this transition itself never signifies that the one fundamental form absolutely disappears while another appears absolutely new in its place." On the contrary:

> The transformation must leave a determinate stock of principles unaffected; for it is undertaken simply for the sake of securing this stock, which [securing] in fact reveals its proper goal. Since we never compare the totality of hypotheses with the naked facts, but can always only oppose one hypothetical system of principles with another, more comprehensive and radical [system], we require for this progressive comparison an ultimately constant *measure* in highest principles that hold for all experience in general. ... In this sense the critical theory of experience hopes in fact to construct the *universal invariant theory of experience*, as it were, and thereby fulfill a demand towards which the inductive procedure itself ever more clearly presses. ... The goal of critical analysis would be attained if it succeeded in establishing in this way what is ultimately common to all possible forms of scientific experience, that is, in conceptually fixing those elements that are preserved in the progress from theory to theory, because they are the conditions of each and every theory. This goal may never be completely attained at any given stage of knowledge; nevertheless, it remains as a *demand* and determines a fixed direction in the continual unfolding and development of the system of experience itself.[79]

As Cassirer here suggests, not only are we never in a position to take our

---

79. Cassirer (1910/23, pp. 355–7/pp. 268–9).

present constitutive principles as ultimate, we are also never in a position to know how the future evolution of constitutive principles will actually unfold. The best we can do, at any given time, is make an educated guess, as it were, as to what these ultimate, maximally general and adequate constitutive principles might be.[80] (At the present time, for example, it would seem that both infinitesimally Euclidean geometry and principles of natural motion essentially tied to the underlying geometry of space-time are very good bets.)

This last point leaves us with an important question, however. How is it possible to venture a transformation of our present constitutive principles resulting in a genuine conceptual change or shift in paradigm? How, more specifically, can the proposal of a radically new conceptual framework be, nonetheless, both rational and responsible? In accordance with our threefold perspective on inter-paradigm convergence we can now say the following: first, that the new conceptual framework or paradigm should contain the previous constitutive framework as an approximate limiting case, holding in precisely defined special conditions;[81] second, that the new constitutive principles should also evolve continuously out of the old constitutive principles, by a series of natural transformations; and third, that this process of continuous conceptual transformation should be motivated and sustained by an appropriate new philosophical meta-framework, which, in particular, interacts productively with both older philosophical meta-frameworks and new developments taking place in the sciences themselves. This new philosophical meta-framework thereby helps to define what we mean, at this point, by a natural, reasonable, or responsible conceptual transformation.

In the transition from Aristotelian-Scholastic natural philosophy to classical mathematical physics, for example, at the same time that Galileo was subjecting the Aristotelian conception of natural motion to a deep (yet continuous) conceptual transformation, it was also necessary to eliminate the hierarchical and teleological elements of the Aristotelian conceptual frame-

---

80. As a matter of fact, Cassirer (and the Marburg School more generally) does not defend a relativized conception of a priori principles. Rather, what is *absolutely* a priori are simply those principles that remain throughout the ideal limiting process. In this sense, as explained in Friedman (2000b, chapter 7), Cassirer's conception of the a priori is purely regulative, with no remaining constitutive elements. The present conception results from combining the relativized yet still constitutive a priori developed within the logical empiricist tradition with the Marburg version of the regulative use of reason. (Here I am particularly indebted to comments from Henry Allison.)

81. It is important to note, however, that this limiting procedure does not (retrospectively) preserve constitutivity: see again the discussion in Part Two, section 3 below, especially the two full paragraphs to which note 37 is appended.

work in favor of an exclusively mathematical and geometrical point of view – which was encapsulated, at the time, in the distinction between primary and secondary qualities. Euclidean geometry, as an exemplar of rational inquiry, was of course already a part of the Aristotelian framework, and the problem then was, accordingly, to emphasize this part at the expense of the hylomorphic and teleological conceptual scheme characteristic of Aristotelian metaphysics. This task, however, required a parallel reorganization of the wider concepts of Aristotelian metaphysics (concepts of substance, force, motion, matter, mind, space, time, creation, divinity), and it fell to the philosophy of Descartes to undertake such a reorganization – a philosophy which in turn interacted productively with previous scientific advances such as Copernican astronomy, new results in geometrical optics, and Descartes's own invention of analytic geometry. Similarly, in the transition from classical to relativistic mechanics (both special and general), at the same time that Einstein was subjecting the classical conceptions of space, time, motion, and interaction to a deep (yet continuous) conceptual transformation, philosophical debate surrounding the foundations of geometry, namely, that between Helmholtz and Poincaré, in which empiricist and conventionalist interpretations of that science opposed one another against the ever present backdrop of the Kantian philosophy, played an indispensable role – and, in turn, was itself carried out in response to the new mathematical advances in the foundations of geometry made throughout the nineteenth century. It takes, to be sure, the genius of a Descartes, a Newton, or an Einstein to interweave philosophical meta-frameworks with scientific paradigms productively in this way. Yet, as these examples clearly show, it can in fact be done.[82]

The present conception of a convergent sequence of successive frameworks or paradigms, approximating in the limit (but never actually reaching) an ideal state of maximally comprehensive communicative rationality in which all participants in the ideal community of inquiry agree on a common set of truly universal, trans-historical constitutive principles, need not imply a second and further conception of convergence, according to which successive scientific theories are viewed as ever better approximations to a radically external world existing entirely independently of the scientific enterprise itself. The original Kantian conception of objectivity, in particular, was explicitly intended to undermine such a naively realistic interpretation of scientific knowledge, through its sharp distinction between appearances and

---

82. Again, this entire matter is discussed in considerably greater detail in Part Two, section 4 below.

things in themselves, and its accompanying insistence that our best knowledge of nature – natural scientific knowledge – extends only to appearances. So the way is similarly open, on our modified Kantian conception, simply to define scientific truth, in a Peircean vein, as whatever the ideal community of inquiry eventually agrees to. Or, in the even more radical, "logical idealist" tradition of the Marburg School, we might characterize the actual empirical world as that limiting mathematical structure towards which the progress of natural science is in fact converging.[83]

Finally, and by the same token, the present conception of scientific rationality need not imply the elimination of all genuine contingency from scientific progress, in the sense that there is a single pre-ordained route through the set of all possible constitutive principles, as it were, which the evolution of science necessarily follows at each stage. On the contrary, we can, if we like, imagine a branching tree structure at every point, so that alternative future evolutions of our fundamental constitutive principles are always possible. Scientific rationality, from the present point of view, does not require the existence of a single fixed route through this structure, any more than it requires a single set of constitutive principles fixed once and for all. What we require, rather, is that *any* reasonable route through the tree be convergent, for it is this, and no more, that is implied by the notion that our present scientific community is an approximation to an ideal limiting community of inquiry acknowledging similarly ideal standards of universal, trans-historical communicative rationality. The aim of our modified version of Kantianism, as we pointed out at the end of Lecture II, has nothing to do with certainty or epistemic security at all. It aims, rather, at precisely such universal rationality, as our reason grows increasingly self-conscious and thereby takes responsibility for itself.

---

83. The present conception of scientific rationality is *consistent* with such "anti-realist" conceptions of scientific truth, but it is in no way committed to them. Perhaps the best way to put the point, as I do in Part Two, section 5 below, is that the present conception of scientific *rationality* does not involve a parallel conception of scientific *truth* – either "realist" or "anti-realist." Thus, for example, as I observe in Part Two, note 58 below, the convergence required by a Peircean account of scientific truth is actually much stronger than that required by our own account of scientific rationality (and see the remarks on convergence and rationality immediately below).

PART TWO

Fruits of Discussion

1

The Relativized A Priori

The notion of a relativized yet still constitutive a priori, as developed especially in Lecture II, is central to the modified version of a Kantian philosophy of science I am attempting to articulate. The idea is that advanced theories in mathematical physics, such as Newtonian mechanics and Einsteinian relativity theory, should be viewed as consisting of two asymmetrically functioning parts: a properly empirical part containing laws such as universal gravitation, Maxwell's equations of electromagnetism, or Einstein's equations for the gravitational field; and a constitutively a priori part containing both the relevant mathematical principles used in formulating the theory (Euclidean geometry, the geometry of Minkowski space-time, the Riemannian theory of manifolds) and certain particularly fundamental physical principles (the Newtonian laws of motion, the light principle, the equivalence principle). Although we explicitly acknowledge that what we are here calling a priori principles (both mathematical and physical) change and develop along with the continual progress of empirical natural science, and in response to empirical findings, we still insist, against Quinean epistemological holism, that these principles should nonetheless be seen as constitutively a priori in something very like the original Kantian sense.

This idea is of course *prima facie* quite puzzling in our post-Quinean philosophical environment. What can it possibly mean to call principles a priori that change and develop in response to empirical findings? According to the traditional conception of the a priori, in which it means "justified independently of experience," it would seem that any principle correctly characterized as a priori would perforce have to hold (if it does hold) entirely independently of all empirical findings and would thus have to hold "come what may." Once we grant (in the context of truly revolutionary scientific developments) that principles of geometry and mechanics, say, are by no means unrevisable, then it would appear that there is no sense left in calling such principles a priori. To be sure, in perfect accordance with Duhemian-Quinean holism, we *can* hold such principles come what may if, for whatever reason, we then decide to do so. But this type of unrevisability or independence from experience applies equally to all principles of natural science, including those we are here calling properly empirical laws.

The first point to make, in response to this kind of objection, is that the concept of the relativized a priori, as originally formulated within the tradition of logical empiricism, was explicitly intended to prise apart two meanings that were discerned within the original Kantian conception: necessary and unrevisable, true for all time, on the one hand, and "constitutive of the concept of the object of [scientific] knowledge," on the other. As we pointed out in Lecture II, such an explicit separation was fundamental to both Reichenbach's initial formulation of the idea in his *Theory of Relativity and A Priori Knowledge* (1920) and Carnap's later generalization of it in his philosophy of formal languages or linguistic frameworks. Indeed, in section 82 of his *Logical Syntax of Language* (1934), Carnap emphatically asserts that any sentence of the language of mathematical physics, including the L-rules or analytic sentences, may be revised in light of a "recalcitrant" protocol-sentence, and he also explicitly embraces Duhemian holism in the empirical testing of synthetic sentences. Nevertheless, the L-rules, in sharp contrast with the P-rules, *define* what it means for a protocol-sentence to stand in logical relations to a synthetic sentence in the first place, so that empirical testing – however holistic – then has a precise logical meaning. In this sense, analytic sentences or L-rules cannot themselves be logically *tested* by experience, although, on purely pragmatic grounds, they can be revised in response experience.[1]

---

1. See Carnap (1934/37, section 82, pp. 245–6/pp. 317–9): "A sentence of physics, whether it is a P-fundamental sentence or an otherwise valid sentence or an indeterminate assumption (i.e., a premise whose consequences are investigated), is *tested*, in that consequences are deduced from it on the basis of the transformation rules of the language until one finally arrives at propositions of the form of protocol-sentences. These are compared with the protocol-sentences actually accepted and either confirmed or disconfirmed by them. If a sentence that is an L-consequence of certain P-fundamental sentences contradicts a proposition accepted as a protocol-sentence, then some alteration must be undertaken in the system. ... No rule of the physical language is definitively secured; all rules are laid down only with the proviso that they may be altered as soon as it seems expedient. That holds not only for the P-rules, but also for the L-rules including mathematics. In this respect there is only a graduated distinction: in the case of some rules it is more difficult to decide to give them up than it is for others. [If we assume that a newly appearing protocol-sentence within the language is always synthetic, then there is nonetheless the following difference between an L-valid and thus analytic sentence S_1 and a P-valid sentence S_2, namely, that such a new protocol-sentence – whether or not it is acknowledged as valid – can be at most L-incompatible with S_2 but never with S_1. Nevertheless, it can happen that, under the induce-

This separation of two meanings of the a priori is prefigured within the original Kantian conception by Kant's own novel understanding of (synthetic) a priori truth. Although Kant of course took it for granted that principles he viewed as a priori (basic principles of geometry and mechanics, for example) were necessary, unrevisable, apodictally uncertain, and so on, his characteristic explanation of the possibility of such principles rather highlighted their constitutive function with respect to a posteriori or empirical truths – the function of making the empirical cognition of such truths (which, for Kant, includes what we would now call their empirical confirmation) first possible. The reason that a priori knowledge is in fact independent of empirical cognition or experience, for Kant, is that a priori knowledge yields the necessary conditions under which alone empirical cognition or experience can take place.[2] Since they formulate the necessary conditions or rules for establishing empirical knowledge, a priori principles cannot themselves be similarly established; and it is in precisely this sense that they are prior to or independent of experience.[3] Once we acknowledge, as we must, that principles Kant took to be a priori can after all be revised, the way is then open, as it was for Reichenbach and Carnap, to retain Kant's characteristic understanding of a priori principles as constitutive in this sense while rejecting the more traditional marks of necessity, unrevisability, and apodictic certainty.

It is still not entirely clear, however, what exactly it means for (putatively) a priori principles to be necessary conditions of empirical knowledge. And it is perfectly understandable, in particular, if the reader still has doubts whether a Kantian conception of this constitutive relationship can be maintained in the face of the revolutionary scientific developments that have led to our post-Quinean philosophical predicament. What I want to argue, in response to such doubts, is that the revolutionary developments in question have made the constitutive relationship between a priori and empirical principles even

---

ment of new protocol-sentences, the language is so altered that S_1 is no longer analytic.]" (The translation inadvertently has "incompatible" rather than "L-incompatible" in the penultimate sentence.) For further discussion see Friedman (1999, chapter 9, section IV). As we point out in note 15 below, however, this Carnapian way of articulating the distinction does in fact ultimately lead to Quinean holism.

2. See A146/B185 in the Schematism chapter of the *Critique of Pure Reason*: "But all our cognitions lie in the whole of all possible experience, and in the universal relation to this consists that transcendental truth which precedes all empirical truth and makes it possible."

3. For an extended discussion of Kant's characteristic conception of the constitutive function of a priori principles see again De Pierris (1993). De Pierris takes pains, in particular, to contrast Kant's conception with the rationalist understanding of "clear and distinct ideas."

clearer and more explicit than it was before – and they have, accordingly, in fact undermined decisively the contrary position of thoroughgoing epistemological holism.

We have said that constitutive principles are necessary conditions of the possibility of properly empirical laws. But this does not mean that they are necessary conditions in the standard sense, where A is a necessary condition of B simply if B implies A. To say that A is a constitutive condition of B rather means that A is a necessary condition, not simply of the truth of B, but of B's meaningfulness or possession of a truth value. It means, in now relatively familiar terminology, that A is a *presupposition* of B. Thus, in the well-worn example originally due to Russell, "The present King of France is bald" presupposes that there is one and only one present King of France, in the sense that the proposition in question lacks a truth value if its accompanying presupposition does not hold.[4] Similarly, in our example from Newtonian physics, the law of universal gravitation essentially employs a concept – absolute acceleration – which has no empirical meaning or application (within the context of Newtonian physics) unless the laws of motion hold. Within the context of Newtonian physics, that is, the only way in which we know how to give empirical meaning and application to the law of universal gravitation is by presupposing that the laws of motion are true: if the latter principles are not true (in the sense that there exists a frame of reference in which they hold) then the question of the empirical truth (or falsity) of the law of universal gravitation cannot even arise.

Yet the mere idea of a presupposition in this sense is of course much too weak to capture the Kantian notion we are after. For we certainly do not want to say that every presupposition of a properly empirical statement is thereby constitutively a priori: think of the present King of France. We want to reserve this characterization for particularly fundamental presuppositions lying at the basis of mathematical physics – principles which, accordingly, can plausibly be taken as fundamental presuppositions of *all* empirical truth (at least in the natural sciences). The privileged position of these particular presuppositions is then doubtless due, to a large extent, to precisely their fundamental

---

4. This kind of interpretation of Kantian necessary or constitutive conditions is developed in Brittan (1978, pp. 28–42), relying on van Fraassen's semantical account of the notion of presupposition (and, in the particular case of denoting phrases, on Strawson's anti-Russellian account of such phrases). As Brittan also points out, writers in the pragmatist tradition sometimes use the bare logical notion of necessary condition (as simple implication) to reduce the notion of presupposition to absurdity. Brittan cites Pap (1946, p. 1) – which is explicitly indebted to the discussion in Lewis (1929, pp. 197–202).

character, to their extreme generality with respect to the totality of empirical truth. But it is also due, I now want to suggest, to a specific problem involving the application of the concepts of space, time, and motion that is characteristic of modern mathematical physics – a problem that becomes ever more explicit with the development of relativity theory.

Thus, in pre-modern, Aristotelian-Scholastic physics, there is no particular difficulty about applying the concepts of space, time, and motion to experience. Space consists of a three dimensional Euclidean sphere whose center is occupied by the earth and whose bounding surface is occupied by the fixed stars. The entire celestial sphere uniformly rotates daily from east to west, and smaller concentric celestial spheres, each bearing an accompanying heavenly body, uniformly rotate in the contrary direction with considerably longer periods (the sun yearly, the moon monthly, and so on). Each heavenly body thus has a natural place determined by its celestial sphere and a natural motion determined by the rotation of this sphere. (I here gloss over the problem of the planets.) Moreover, in the sublunary or terrestrial region contained within the innermost celestial sphere centered on the earth, the four elements (earth, water, air, and fire) each have their natural places ranging concentrically from the center to the periphery, as well as a characteristic state of natural motion in a straight line from whatever unnatural place in which they might find themselves back to their natural places. The point, in our present context, is that our theoretical concepts of space, time, and motion fit together smoothly and unproblematically with the world as it present itself to sense experience: the "scientific image" is a straightforward systematization of the "manifest image."

By contrast, in the modern mathematical physics of the sixteenth and seventeenth centuries that culminated in the work of Newton, this unproblematic fit between physical theory and sense experience is irretrievably lost. Instead of a finite three dimensional Euclidean sphere, space now consists of the whole of three dimensional Euclidean extension – infinite in all directions – which, for this very reason, no longer contains any privileged positions at all. Since no particular body is necessarily at the "center" of this space (there is no such intrinsic center), we can construct an equivalent Euclidean space centered on any physical body we wish (the earth, the sun, and so on). We can distinguish, that is, an infinite variety of *relative spaces* (reference frames), none of which has so far been privileged. As a result, our new fundamental concept of natural motion – given by the law of inertia – is also essentially ambiguous. Bodies acted on by no external forces uniformly traverse Euclidean straight lines to infinity; but relative to *which* relative space(s) is this supposed to hold? Finally, since the uniform passage of time is no longer directly given by any observable physical motion (such as the diurnal rotation of the fixed stars), but rather by a natural state of inertial motion that is never actually observed, the notion of time is essentially ambiguous in a precisely parallel sense. Neither space nor time nor motion has an unequivocal relation to our experience.

One way to describe this fundamental problem is that the mathematical representations employed in modern physics have become increasingly abstract in relation to concrete sensory experience. Infinite Newtonian space is not sensibly given like finite Aristotelian space – nor is natural inertial motion given like natural Aristotelian motion, uniform Newtonian time like uniform Aristotelian time. For precisely this reason, however, there is a new problem of somehow *coordinating* our new mathematical representations with concrete sensible experience before we are even in a position to be fully explicit about what our new physical theory actually says. This problem (which, at bottom, is the characteristically modern problem of absolute versus relative motion) was not definitively solved until Newton formulated the laws of motion in the *Principia*. For, as we now understand it, these laws of motion define a privileged class of relative spaces or reference frames (what we now call inertial frames) in which the modern concepts of space, time, and motion then unambiguously apply.[5] Thus we can, as in Book III of the *Principia*, empirically establish that the center of mass of the solar system determines such an inertial frame to a very high degree of approximation, take the motions defined in this frame to be absolute motions (again to a very high degree of approximation), and approximate the uniform passage of time by these same empirically determinable motions (correcting the diurnal rotation of the earth, for example, in light of tidal friction).

The laws of motion, in the context of Newtonian physics, therefore function as what Reichenbach, in his 1920 book on relativity and the a priori, aptly calls *coordinating principles* (axioms of coordination). They serve as general rules for setting up a coordination or correspondence between the abstract mathematical representations lying at the basis of Newtonian physics (infinite

---

5. This way of viewing the laws of motion was not fully clarified until the concept of inertial frame was explicitly characterized in the late nineteenth century in the work of Carl Neumann, James Thomson, and Ludwig Lange. For discussion see Torretti (1983, section 1.5), and especially DiSalle (1991). From this point of view (and from a Kantian point of view), the metaphysical obscurity surrounding Newton's original notion of absolute space arises from insufficient clarity about the fundamentally new problem of coordinating abstract mathematical structures with concrete sensible experience created by modern physics.

Euclidean space, uniformly traversed straight lines in this space, abstract temporal intervals during which such states of uniform motion traverse equal spatial intervals) and concrete empirical phenomena to which these representations are intended to apply (the observable relative motions in the solar system, for example).[6] And without such general rules of coordination we simply have no idea what it means for concrete empirical phenomena to be described by the mathematical representations in question – either correctly or incorrectly.[7] The Newtonian laws of motion are thus presuppositions of the properly empirical laws of Newtonian physics (such as the law of gravitation) in the sense considered earlier, but they are also presuppositions of a very special sort. Their peculiar function is precisely to mediate between abstract mathematical representations and the concrete empirical phenomena these abstract mathematical representations are intended to describe. As such, they do in fact fulfill the characteristically constitutive function first delimited by Kant, and, accordingly, they have a genuine claim to be thereby considered as constitutively a priori.

Now, as we know, the particular coordinating principles employed in Newtonian physics have been radically modified in the further development of mathematical theories of space, time, and motion. Indeed, the mathematical representations lying at the basis of these theories have themselves become increasingly abstract. In place of an infinite, three dimensional Euclidean space we now use a four dimensional, (semi-)Riemannian manifold of variable curvature – a four dimensional manifold endowed with a metric of Lorentzian signature, so that, in effect, a "light cone" is defined at each point of the manifold mimicking infinitesimally the (flat) space-time geometry of special relativity. And, in place of the inertial trajectories of Newtonian physics (which, in an important sense, are retained virtually unchanged in special relativity) we now distinguish the four dimensional

---

6. Such general coordinating principles should be distinguished from *operational definitions*, which proceed by coordinating some actual concrete phenomenon with an abstract theoretical concept. Thus, if we define the uniform passage of time by stipulating that some actual periodic process (such as the diurnal rotation of the earth) is uniform, this would be an operational definition of "equal times." Defining "equal times" by the laws of motion, by contrast, explicitly provides for the possibility of correcting and refining any and all such concrete coordinations without limit.

7. It is in this sense that the new problem of coordinating mathematical representations with experience is entirely different from the traditional Platonic problem, which rather concerns the circumstance that exact mathematical representations are only *approximately* realized in sense experience. The new problem concerns what it means for mathematical representations to apply to sense experience at all – whether correctly or incorrectly, exactly or approximately.

geodesics of the (semi-)Riemannian metric as representing our new fundamental state of natural motion. Finally, in place of the Newtonian law of
universal gravitation, formulated using the mathematical representations of
the old constitutive framework, we now use Einstein's field equations governing the four dimensional space-time metric, equations which relate this
metric to our new mathematical representation of matter (the so-called
stress-energy tensor).

The abstract mathematical representations lying at the basis of this new
theory of space, time, and motion – Einstein's general theory of relativity –
are even more radically disassociated from sense experience than those of
Newtonian physics. For infinite three dimensional Euclidean space can still
plausibly be taken as a fundamentally intuitive "form" of our human sense-
perception (as it was for Kant), even though, as we have seen, the application
of this representation to concrete physical phenomena becomes profoundly
ambiguous in modern mathematical physics. But the four dimensional, vari-
ably-curved geometry of general relativity is an entirely non-intuitive
representation having no intrinsic connection whatever to ordinary human
sense experience. And it is for this reason, above all, that both the logical
empiricists and Einstein himself discern an intimate and essential relationship between the general theory of relativity, on the one hand, and the
modern "formal" or "axiomatic" conception of geometry associated with
David Hilbert, on the other. For the entire point of the modern axiomatic
conception of mathematics, on this view, is to overturn the connection
between mathematics and sensory experience once and for all, leaving us with
a radically new view of mathematics (including geometry) as dealing solely
with pure abstract relational structures.[8]

On this new view of mathematics there is thus more need than ever for
principles of coordination to mediate between abstract mathematical structures and concrete physical phenomena. So it is in no way accidental that
coordination as a philosophical problem was first articulated by scientific
philosophers deliberately attempting to come to terms with Einstein's general
theory of relativity. Indeed, Reichenbach in 1920, together with Moritz
Schlick in virtually contemporaneous work, were the first thinkers explicitly
to pose and to attempt to solve this philosophical problem. And the solution

---

8. For details of these developments see Friedman (2001). The story is actually a rather complex
one, involving, in particular, an intermediate stage of nineteenth century work on the foundations of geometry (by Helmholtz and Poincaré especially) where the non-Euclidean spaces of
constant curvature are taken to generalize the essentially perceptual conception of geometry
elaborated by Kant. See also section 4 below.

at which they both arrived is that there is a special class of non-empirical physical principles – variously called coordinating or constitutive principles by Reichenbach, conventions in the sense of Henri Poincaré by Schlick – whose function is precisely to establish and secure the required connection between abstract mathematical structures and concrete sensory experience.[9] In the case of the general theory of relativity, in particular, the required coordination is established by two fundamental Einsteinian principles: the light principle and the principle of equivalence. The law of the constancy and source-independence of the velocity of light coordinates concrete physical phenomena with the Lorentzian (or infinitesimally Minkowskian) character of the new four dimensional space-time metric, and the law that freely falling "test particles" in a gravitational field follow four dimensional geodesic paths of this metric then completes the coordination.[10] These two Einsteinian principles are thus the counterparts of the Newtonian laws of motion: they, too, are fundamental mathematical-physical presuppositions without which the properly empirical laws of our new theory (Maxwell's equations in a relativistic context, Einstein's equations for the gravitational field) have no empirical meaning or application at all.[11]

What we end up with, then, is the following general picture of the structure of our mathematical-physical theories of space, time, and motion (a picture that was implicit in the discussion of Lecture II). Each of the theories in question (Newtonian mechanics, special relativity, general relativity) consists of three asymmetrically functioning parts: a mathematical part, a mechanical part, and a (properly) physical or empirical part. The mathematical part contains the basic mathematical theories, representations, or structures intended

---

9. For details see again Friedman (2001). For the disagreement between Reichenbach and Schlick concerning "constitution" versus "convention" see also Friedman (1999, chapter 3). For another perspective on this disagreement–and on the idea of relativized a priori principles more generally–see Parrini (1998).

10. The function of these two principles as the foundation for the empirical content of general relativity is made particularly clear and precise in Ehlers, *et. al.* (1972). Although they establish a more direct coordination with observable phenomena than do the Newtonian laws of motion, the two Einsteinian principles are still not concrete coordinations or operational definitions in the usual sense (see note 6 above). They rather establish norms or ideal cases which actual concrete phenomena can only approximate in the limit: only truly unextended or infinitesimal "test bodies" exactly follow four dimensional geodesics, and the light principle definitive of special relativity similarly holds exactly only in strictly infinitesimal regions.

11. Here, it is especially illuminating to ask oneself what it could possibly mean to assert that "space-time has variable curvature determined by the distribution of mass and energy" in the absence of the principle of equivalence. For an excellent discussion of this point see again DiSalle (1995).

to describe the spatio-temporal framework in question (infinite Euclidean space, four dimensional Minkowski space-time, [semi-]Riemannian space-time manifolds). The physical or empirical part then attempts to use these mathematical representations in formulating precise empirical laws describing some concrete empirical phenomena (the law of universal gravitation, Maxwell's equations for the electro-magnetic field, Einstein's equations for the gravitational field). In order to accomplish this, however, we need principles of coordination comprising the mechanical part (the Newtonian laws of motion, the light principle, the principle of equivalence) whose function is to set up a general correspondence between the mathematical part, on the one side, and concrete empirical phenomena, on the other, in such a way that the precise laws of nature formulated with the help of the mathematical part in fact have empirical meaning.

Given such a tripartite structure, the laws of nature comprising the (properly) physical part can then be empirically tested: for example, by Newton's description of the solar system (including planetary perturbations) in *Principia*, Book III, or Einstein's calculation of the advance of the perihelion of Mercury. It is a profound mistake of Quinean holism, however, to view this procedure as empirically testing the other two parts in the same way. For, in the first place, it is clear that the mathematical part of our theories, considered independently of the empirical application in question, is in no way empirically tested by such a procedure; what is empirically tested is rather the particular coordination or correspondence in virtue of which some or another mathematical structure is used to formulate precise empirical laws about some or another empirical phenomena. Thus, for example, the theory of Riemannian manifolds is used in formulating both general relativity and versions of classical mechanics (in Hamiltonian formulation).[12] But the theory of Riemannian manifolds itself is empirically tested by neither of these applications (each of which employs its own characteristic coordination); it remains a purely abstract description of certain mathematical structures, whose distinctive theorems and principles are justified purely mathematically.

---

12. In these formulations of classical mechanics our underlying manifold is configuration space rather than physical space or space-time; a Riemannian metric on the manifold represents kinetic energy rather than geometrical distance; and the function of our metric is to induce a one-parameter group of automorphisms (representing temporal evolution) on the phase space of the system generated (infinitesimally) by the Hamiltonian function (kinetic plus potential energy). This way of formulating classical mechanics thus facilitates a comparison with quantum mechanics. As Lützen (1995) shows in detail, these formulations can be traced back to nineteenth century work on Hamiltonian mechanics, especially to work of Rudolf Lipschitz directly following upon that of Riemann.

Similarly, and in the second place, it is also a mistake to view the coordinating principles comprising the mechanical part as being empirically tested by the same procedure. For, as we have emphasized repeatedly, the procedure in question could not even be set up in the first place without some or another coordinating principle already in place. Thus, for example, it is not that Einstein's field equations cannot be empirically tested in the absence of the principle of equivalence for Duhemian reasons: it is not as if we were faced with two empirical assertions jointly functioning to generate empirical predictions (such as the conjunction of an empirical law with a description of a measuring device or experimental apparatus used in testing that law). Rather, in the absence of the principle of equivalence, Einstein's field equations remain a purely mathematical description of a class of abstract (semi-)Riemannian manifolds with no empirical meaning or application whatsoever.[13]

The increasingly abstract character of the mathematics applied in our most fundamental physical theories, the characteristically modern problem of coordination between abstract mathematical structures and concrete physical phenomena, and the Kantian idea of constitutively a priori principles functioning to mediate between the two are thus three distinguishable aspects of what is at bottom the same conceptual situation. And it is only by ignoring this situation entirely that we can, on the contrary, arrive at Quinean holism. Indeed, Quine himself arrives at epistemological holism by focussing exclusively on problems in the foundations of mathematics, with no real concern for the foundations of modern mathematical physics. Quine's basic move in response to the failure of traditional logicism is decisively to reject the inclusion of set-theory (viewed as our most general theory of what I am here calling abstract mathematical structures) within logic: logic is identified with first-order or elementary logic, while set-theory, with its heavily "platonistic" existential commitments, is identified with mathematics as distinct from logic. Yet these same existential commitments give rise to ontological and epistemological scruples deriving from an underlying philosophical outlook sympathetic to traditional nominalism and empiricism – so that, in particular, when Quine acknowledges that "constructive nominalism" is a failure, his characteristic form of epistemological holism then presents itself as an attractive empiricist alternative. For we can now view our total system of natural science as a conjunction of set-theory with various scientific theories ordi-

---

13. This is not to say, however, that coordinating principles have no empirical content at all or that empirical testing is entirely irrelevant to them (as it is in the case of pure mathematics). This matter will be further addressed in the next section.

narily so-called, a system which is tested as a whole by the deduction (in first-order logic) of various empirical consequences within this total system. The ontological commitments of set-theory are thus empirically justified to the same extent, and in the same way, as are our "posits" of any other theoretical entities in natural science.[14] The present discussion has made it clear, I hope, that, whatever motivations it might have in the context of recent debates in the foundations of mathematics, this view serves only to obscure the situation entirely in the (complementary) foundations of modern mathematical physics.[15]

---

14. This general line of thought, taking its starting point from the acknowledged failure of Goodman and Quine (1947), is initiated in Quine (1948/53) and plays a central role in Quine's philosophy throughout his career: see, e.g., Quine (1955/66, section VI), (1960, § 55).

15. It must be admitted, however, that a considerable share of the responsibility for this obscurity must be assigned to Carnap himself; for the problem of coordination prominent in the early work of Schlick and Reichenbach eventually becomes lost in Carnap's philosophy of linguistic frameworks. The full story is a complex one, but the point, briefly, is that Carnap is led, in the *Aufbau*, to incorporate the problem of coordination *within* the logical systematization of science he constructs there. Instead of a problem relating abstract logico-mathematical systems to concrete empirical reality, we now have a problem, within a particular logico-mathematical system, of distinguishing logical and descriptive terms, analytic and synthetic sentence. This then becomes the program of *Logical Syntax*, which is motivated, above all, by problems in the foundations of mathematics. [Aspects of this complex story are discussed in Friedman (1999, especially the Postscripts to chapters 1 and 6).] So when Quine rejects Carnap's analytic/synthetic distinction we thereby end up with epistemological holism (compare note 1 above).

2

A Priori Principles and Empirical Evidence

On the view of theories in mathematical physics outlined above, the role of what I am calling constitutively a priori principles is to provide the necessary framework within which the testing of properly empirical laws is then possible. Without a constitutive framework, the putatively empirical laws would have no empirical content after all, but would simply belong to the domain of pure mathematics. With a constitutive framework already in place, however, properly empirical laws can be confronted with sensory experience and the empirical world in a particularly clear and direct fashion: one can compare calculated values of various physical magnitudes and parameters (such as the rate of precession of the equinoxes, say, or the advance of the perihelion of Mercury) with actually observed and measured values and thereby obtain an exact quantitative estimate of the fit between theory and experience. What makes such a calculation count as *evidence* either for or against the theory (more precisely, for or against the properly empirical laws of the theory) is a prior acceptance of the constitutive framework that secures the empirical content of the theory.

This view of empirical evidence differs from a purely pragmatic or instrumental conception of theory testing, according to which theories are essentially instruments for accurate prediction and are acceptable or not to the extent to which they succeed in this pragmatic task. Both a Newtonian and an Einsteinian physicist, for example, can and must agree that general relativity yields more accurate predictions for the advance of the perihelion of Mercury. From the Newtonian physicist's point of view, however, general relativity can *only* be accepted as a pragmatically acceptable device for prediction; it *cannot* be a true description of empirical reality. For, from the Newtonian point of view, the constitutive framework of general relativity is not even possible or coherent, and there is thus no sense in which Einstein's field equations can actually be empirically true. Only when the constitutive framework of general relativity (the Riemannian theory of manifolds, the light principle, the principle of equivalence) is already in place is it possible for the field equations to be empirically true; and it is only within the context of this already accepted framework, therefore, that Einstein's calculation of the perihelion of Mercury can then count as genuine empirical evidence – again either for or against.[16]

A constitutive framework thus defines a space of empirical possibilities (statements that can be empirically true or false), and the procedure of empirical testing against the background of such a framework then functions to indicate which empirical possibilities are actually realized. Moreover, just as in the original Kantian conception of the a priori, the relevant notion of possibility here has two distinguishable aspects. On the one hand, there is what we might call a purely logical notion of possibility given by the mathematical part of our constitutive framework. Without the Riemannian theory of manifolds, we might say, the space-time structure of general relativity is not even logically possible, and so, *a fortiori*, it is empirically impossible as well. On the other hand, however, mere logical possibility in this sense is clearly not sufficient for empirical possibility. We need, in addition, the coordinating principles of our theory (the light principle and the principle of equivalence), and it is precisely these principles, then, that define what we might call *real* (as opposed to merely logical) possibility. Einstein's field equations are thus logically possible as soon as we have Riemannian manifolds available within pure mathematics, but they are only really possible (possible as an actual description of some empirical phenomena) when these abstract mathematical structures have been successfully coordinated with some or another empirical reality.[17]

---

16. Here I am especially indebted to discussions at Göttingen with Felix Mühlhölzer. Note that this issue is quite distinct from the traditional problem of realism versus instrumentalism. The difficulty the Newtonian physicist has with the space-time structure of general relativity, for example, has nothing directly to do with worries about unobservable entities. The problem is rather that the space-time structure in question is not even a coherent possibility: whether or not it can in any sense be "observed," it cannot be coherently *described*. Thus the "constructive empiricism" defended by van Fraassen (1980), which is explicitly distinguished from traditional instrumentalism, is, as far as I can see, entirely compatible with the conception of empirical evidence outlined here.

17. For the Kantian notion of real possibility see, e.g., the *Critique of Pure Reason* at Bxxvi,n. It is precisely because we retain a counterpart to this notion that the Kantian terminology of constitutive principles is still appropriate. Of course there are also quite fundamental differences between the present notion and the original Kantian one – most notably, Kant's conception of pure intuition, and of the "schematism" of the pure concepts of the understanding therein, implies that the understanding can a priori impose a *unique* (and therefore fixed) set of constitutive principle on experience. (Here I am particularly indebted to comments from and discussions with Carol Voeller.)

In logic a space of possibilities is associated (as it were by duality) with what Wilfrid Sellars calls a logical space of reasons: a network of logical rela-tionships between the logical possibilities that defines what can count as a logical reason or justification for any such logical possibility.[18] Similarly, in the case of the broader notion of empirical possibility developed here (logical plus real possibility), a constitutive framework for a mathematical-physical theory gives rise to what we might call an empirical space of reasons: a net-work of inferential evidential relationships, generated by both logical-mathematical principles and physical coordinating principles, that defines what can count as an empirical reason or justification for any given real possi-bility. It is precisely in virtue of such a network of already accepted inferential relationships, for example, generated by both the Riemannian theory of man-ifolds and the Einsteinian coordinating principles, that the advance of the perihelion of Mercury can then count as an empirical reason or justification for accepting Einstein's field equations – which is essentially stronger, as we have seen, than a merely pragmatic or instrumental reason for employing the general theory of relativity as a "black box" predictive device.

This distinction between logical and real possibility – as a counterpart to the original Kantian notion of constitutivity (note 17 above) – helps to illumi-nate the sense in which our conception of empirical testing and evidence is also essentially stronger than the traditional hypothetico-deductive account. For the hypothetico-deductive account appeals only to inferential relations defined by formal logic; and, unless special measures are employed, it is therefore easily vulnerable to Duhemian and Quinean holism. Theoretical propositions relate to the empirical evidence they logically imply simply as parts of a logical conjunction, and the evidence in question can then only be viewed as a test of this conjunction as a whole.[19] In our present conception, by contrast, physical coordinating principles are just as much essential parts of the network of empirical evidential relationships as are the principles of logic and mathematics. An empirical test of Einstein's field equations by the advance of the perihelion of Mercury therefore counts as an empirical reason,

---

18. For this terminology see, e.g., Sellars (1956, section 36), which is explicitly indebted to Witt-genstein's notion of "logical space" developed in the *Tractatus*. The Sellarsian notion of a "space of reasons" has recently been given prominence in McDowell (1994). For my own attempt to come to terms with this latter work see Friedman (1996).

19. A purely logical view of evidential inferential relationships can thereby easily lead one into Quinean holism. This process, as we have indicated, begins with Carnap's *Logical Syntax*, where the "special measures" I refer to above are just the Carnapian distinctions between L-rules and P-rules, analytic and synthetic sentences: see notes 1 and 14 above.

in the present sense, for accepting these field equations as empirically true, but it does not count, in the same sense, as an empirical reason for accepting the principle of equivalence. Indeed, as we have seen, if the principle of equivalence is not already accepted prior to this procedure, it cannot generate empirical evidence for Einstein's equations at all – which rather, without the principle of equivalence, revert to being merely logically possible.

It may seem, however, that the principle of equivalence can itself be empirically tested. In particular, it appears to be straightforwardly tested by the well-known experiments of Lorand von Eötvös (1889, 1922), which compare the accelerations due to gravity on plumb bobs of different materials – with the result that these accelerations are always essentially the same, in perfect conformity with the demand of the principle of equivalence that all bodies "fall" along the same trajectories in a gravitational field.[20] And, for that matter, it appears that the other great principle of relativistic physics – the light principle – is empirically testable in an analogous fashion. The famous interferometer experiments of Michelson and Morley (1882, 1887), for example, which result in no detectable influence of the motion of the earth on the velocity of light, seem to supply as good an empirical test as can be imagined for the invariance of the velocity of light in different inertial frames and thus for the light principle as it is first introduced in the special theory of relativity. If these two relativistic principles, which I am especially singling out as constitutively a priori, can thus be empirically tested, then what real point is served by sharply distinguishing, as I want to do, between principles of this type and properly empirical laws? What real point is served by continuing to characterize such principles as a priori?

These questions obviously raise fundamental issues for the view of a priori principles I am here attempting to develop. They also involve fundamental issues concerning the empirical testability of, and empirical evidence for, Einstein's theory of relativity. It will be well worth our while, then, to consider them in some detail. If I am not mistaken, however, it turns out that the present view of dynamical constitutive principles – and, more generally, of scientific conceptual revolutions – is clarified and confirmed in a particularly striking way.

It must certainly be acknowledged, at the outset, that the principles in question do have empirical content. If the Eötvös experiments had detected a

---

20. This brings us back to an issue first broached in note 13 above. For discussion of the Eötvös experiments and other related experiments – explicitly characterized as "tests of the equivalence principle" – see Misner, Thorne, and Wheeler (1973, pp. 13–9, 1054–63). I am here indebted to comments at Princeton from Frank Arntzenius and Gordon Belot.

difference in the accelerations due to gravity in different materials, the principle of equivalence could not simultaneously be maintained. If the Michelson-Morley experiments had not had a null result, the relativistic light principle would perforce be empirically untenable. And something similar must hold, more generally, for *all* coordinating principles: there must always be empirical presuppositions in any such case. Indeed, if this were not so, it would be very hard to see how such principles could possibly serve the function they are supposed to serve, namely, the coordination of some distinguished empirical phenomena to a particular mathematical spatio-temporal structure. A coordinating principle must always have a counterpart in empirical reality, and, if such a counterpart does not exist, the principle is empirically vacuous and thus useless.[21] The crucial question, however, is whether such a principle can thereby become empirically false. Can it be empirically tested and confirmed (or disconfirmed) in the same sense as those principles I am here characterizing as properly empirical laws?

Let us first consider the relativistic light principle. The Michelson-Morley experiments and related experiments do not, strictly speaking, show that light has the same invariant velocity in all inertial frames. They rather show only that there are no detectable effects on the behavior of light due to motion with respect to the aether. And this point is vividly demonstrated, of course, by the circumstance that in the Lorentz-Fitzgerald competitor theory to special relativity the very same empirical fact is incorporated within an essentially classical spatio-temporal structure. Indeed, the Lorentz-Fitzgerald version of the electrodynamics of moving bodies was self-consciously constructed to incorporate the Michelson-Morley experiment in this way (whereas, as far as we know, Einstein's own version was developed entirely independently of this experiment). So the Michelson-Morley experiment can in no way be viewed as an empirical test or "crucial experiment" of special relativity with respect to its theoretical alternatives; and it is not a test, for precisely this reason, of the relativistic light principle.

The problem here is much deeper than in standard cases of empirical underdetermination. The problem is not simply that special relativity and the Lorentz-Fitzgerald theory are empirically equivalent, so that, in addition to

---

21. The Newtonian laws of motion, for example, assert the existence of frames of reference (inertial frames) in which these laws of motion hold. If we are not able empirically to determine such a frame (at least to a high degree of approximation) than the Newtonian constitutive principles simply fail in their coordinating function. Moreover, that the Newtonian laws of motion have such an empirical presupposition is even true on Kant's original understanding of them: see Friedman (1992, chapter 3, section IV, chapter 4, section II).

empirical evidence strictly so-called, methodological principles such as simplicity must be explicitly invoked to settle the matter. The problem is rather that what the Lorentz-Fitzgerald theory takes to be a mere empirical fact – no detectable effects on the behavior of light of inertial motion – is used by Einstein as the basis for a radically new spatio-temporal coordination; for Einstein uses his light principle *empirically to define* a fundamentally new notion of simultaneity and, as a consequence, fundamentally new metrical structures for both space and time (more precisely, for space-time). Whereas Lorentz and Fitzgerald take an essentially classical background structure for space, time, and motion to be already sufficiently well defined and only subsequently locate the new empirical discovery in question as a peculiar (but additional) empirical fact formulated against the background of this classical structure, Einstein calls the whole classical structure into question and uses the very same empirical discovery empirically to define a new fundamental framework for space, time, and motion entirely independently of the classical background. It is in precisely this way, as writers under the influence of Poincaré are fond of putting it, that Einstein has "elevated" an empirical law to the status of a convention – or, as I myself would prefer to put it, to the status of a coordinating or constitutive principle.[22] It is precisely here that an essentially non-empirical element of "decision" must intervene, for what is at issue, above all, is giving a radically new space-time structure a determinate *empirical meaning* – without which it is not even empirically false but simply undefined.

This is not to say, however, that there cannot be empirical motivations for preferring Einstein's new coordination to the former classical one. Indeed, the new empirical discovery in question – undetectability of differences in inertial motion in electrodynamics – provides us with strong empirical motivation, not only for entertaining a new coordination, but also (as Einstein was apparently also the first to see) for doubting the adequacy of the classical coordination. For, if there were in fact an empirical counterpart to the classical notion of absolute simultaneity, then there would be (in the context of elec-

---

22. This idea (as inspired by Poincaré, in particular) of "elevating" an empirical law to the status of a definition or convention lies at the heart of Pap's (1946) conception of the "functional a priori." (Pap takes as his motto, in this connection, a well-known passage from Poincaré (1902/13, p. 165/p. 125).) However, although Pap provides several insightful discussions of what I am here calling coordinating principles (especially with regard to the Newtonian laws of motion), he does not sufficiently distinguish these cases from cases where one simply takes some empirical law (such as Hooke's law of elasticity) to be temporarily fixed as a background for isolating other empirical factors under investigation (such as the elastic limits of various materials). For Pap, in the end, what is functionally a priori is simply what is especially well confirmed or established, and, in this way, Pap's approach is ultimately no different from Quine's appeal to entrenchment.

trodynamics) an empirical counterpart to absolute velocity as well. But the new empirical discovery strongly suggests that there is no such empirical counterpart (otherwise differences in inertial motion would be empirically detectable after all). The classical spatio-temporal structure, which we had assumed in the context of Newtonian physics to be unproblematically empirically well defined, thereby turns out to be empirically meaningless.[23] Here we certainly have an empirical motivation, and a particularly strong one, for preferring the new empirical coordination effected by Einstein. But this situation is not happily likened to more standard cases of empirical underdetermination, where two empirically equivalent hypotheses face off against the background of a common constitutive framework, and methodological principles such as simplicity or conservativeness are then invoked to settle the question.[24] It is not happily viewed, in any sense, as a case of empirical testing.

Let us now turn to the principle of equivalence. This principle was introduced by Einstein after he had already rejected the classical notion of absolute simultaneity in the special theory of relativity. As a consequence, the classical Newtonian theory of gravitation was also untenable, since it explicitly involves instantaneous action at a distance and thus absolute simultaneity (compare note 23 above). Einstein therefore set out to formulate a new theory of gravitation compatible with the new relativistic space-time structure; and he focussed, to begin with, on the already well-known and well-established empirical fact that gravitational and inertial mass are equal, so that all bodies "fall" with the same acceleration in a gravitational field.[25] Einstein then leapt from this well-established empirical fact to the bold "heuristic" principle that

23. In the context of Newtonian physics, the laws of motion appear to give empirical meaning to the classical system of inertial reference frames, and the third law, in particular, provides for an instantaneous action at a distance that could empirically realize absolute simultaneity. Gravitational interaction then appears perfectly to instantiate this notion. The problem, however, is that empirically establishing such (distant) gravitational interactions proceeds via the mediation of light rays (as in observing the motions of the heavenly bodies), eventually raising the very issues about electrodynamics and the velocity of light highlighted in the Michelson-Morley experiments. (I am indebted to discussions with Robert DiSalle for this point.)

24. This kind of case would be illustrated by examples such as Hooke's law and elastic limits discussed by Pap (note 22 above). In cases where Hooke's law appears to be violated, conservativeness and entrenchment would lead us to say that the elastic limit had been exceeded instead. In the example of relativity versus Lorentz-Fitzgerald, by contrast, it is hard to see how methodological principles could help us in any case. Although simplicity might lead us to favor relativity, conservativeness and entrenchment would definitely lead us to favor the latter theory: so here we have simply a methodological stand-off.

25. In this connection, Einstein (1916, section 2) explicitly appeals to Eötvös's (1889) experiment. Of course this property of the gravitational field was already well known in the time of Newton

gravitation and inertia are the very same phenomenon. He proceeded, on this basis, to construct models of gravitational fields from "inertial fields" (generated by non-inertial reference frames wherein inertial forces, such as centrifugal or Coriolis forces, arise) in a special relativistic space-time, and he eventually saw that non-Euclidean geometries are thereby associated with gravitational fields. The final step was to take the four dimensional space-time metric, first introduced by Minkowski in the case of a special relativistic flat space-time, as our representative of the gravitational field, and to describe the variations in the space-time curvature associated with this metric by the Einsteinian field equations. The result was Einstein's general theory of relativity, only completed, after a long struggle, in 1915–16.[26]

Just as in the case of special relativity and the light principle, then, Einstein "elevated" an already accepted empirical fact to the status of a new fundamental coordinating principle. Whereas, in the context of the Newtonian theory of gravitation, the equality of gravitational and inertial mass, and the consequent independence of trajectories in a gravitational field from all properties of the attracted bodies, appeared as an interesting (albeit very important) additional fact subsisting against the background of an already established constitutive framework, Einstein used the very same empirical fact as the basis for a radically new constitutive framework.[27] In particular, within this new space-time framework there is no longer an inertial structure (and thus, in Newtonian physics, a class of inertial reference frames) already pre-existing independently of gravity. On the contrary, the only way in which we can now empirically define an inertial structure in the first place is precisely in terms of the "freely falling" trajectories in a gravitational field. These trajectories now play the role that the (purely hypothetical) trajectories of bodies acted on by no forces at all (inertial trajectories) played in Newtonian

---

and plays an important role in the argument for universal gravitation in *Principia*, Book III (see note 27 below).

26. See Norton (1984/89), (1985/89), Stachel (1980/89). The crucial introduction of non-Euclidean geometry on the basis of the principle of equivalence is further discussed – with reference to Stachel (1980/89), in particular – in section 4 below.

27. Newton formulates a version of what we would now call the principle of equivalence as Corollary VI to the Laws of Motion in *Principia*, Book I, and it plays a crucial role in the argument for universal gravitation in Book III – especially in connection with Proposition VI. For it is in virtue of this property that gravitation generates what Stein (1967) very helpfully calls an "acceleration field." Indeed, this property is so important in Newton's argument that Kant, in his analysis of the foundations of Newtonian physics, appears to count it as belonging to the constitutive framework of Newton's theory along with the laws of motion: for discussion see Friedman (1990), DiSalle (1990). From a modern point of view, however, the inertial structure of classical physics is already determined by the laws of motion, independently of gravitation, and this is the central difference between the Newtonian and Einsteinian theories.

physics, with the all-important difference that they now define a variably curved rather than a flat inertial structure. In this way, gravitational force is directly incorporated into the geometry of space-time and thus into the constitutive framework of our theory.

It follows that the Eötvös experiments and related experiments do not function as empirical tests of the principle of equivalence, although they do certainly provide very strong evidence for the central empirical presupposition of this principle. Gravitational and inertial mass are in fact equal, and all bodies do in fact "fall" the same in a gravitational field. But this empirical fact (already well established and accepted in the time of Newton) does not itself amount to the principle of equivalence as Einstein employs it in general relativity. For it here functions as a fundamental coordinating principle (more precisely, as the empirical counterpart to a coordinating principle), where, just as in the case of the light principle in special relativity, an essentially non-empirical element of "convention" or "decision" must necessarily intervene.

The parallel between this case and the case of the light principle in special relativity becomes even clearer when we note that it is also possible to use the principle of equivalence as the basis for an alternative formulation of the *Newtonian* theory of gravitation. In this formulation (which was discovered subsequently to, and against the background of, Einstein's formulation of the general theory of relativity) we discard the flat inertial structure of traditional Newtonian theory in favor of a new variably curved inertial structure, again based – just as in general relativity – on the "freely falling" trajectories in a gravitational field. Unlike in general relativity, however, we retain the classical notion of absolute simultaneity and, accordingly, the classical metrical structures for both space and time. Therefore, although traditional Newtonian gravitation theory and general relativity are of course not empirically equivalent (they differ, for example, on the advance of the perihelion of Mercury), this new formulation and traditional Newtonian theory are precisely empirically equivalent. In both theories gravitational and inertial mass are equal, and in both theories the "freely falling" trajectories are thereby distinctively singled out from among all others. The difference is that the alternative formulation explicitly uses the "freely falling" trajectories as the basis for a new fundamental coordination of space-time structure, whereas the traditional formulation retains the old coordination based on the class of inertial frames – which are themselves defined by the laws of motion functioning, entirely independently of gravitation, as coordinating or constitutive principles.[28]

---

28. For details on the alternative formulation of Newtonian gravitation theory (which was originally discovered by the mathematician Élie Cartan in 1923–24) see Friedman (1983, sections III.4

The alternative formulation of the Newtonian theory of gravitation puts us in a position to make a final, and rather remarkable observation. For, on the basis of this formulation, we can place both general relativity and the Newtonian theory of gravitation within a common constitutive framework given, essentially, by the Riemannian theory of manifolds and the principle of equivalence. In this framework we can now compare both the Newtonian theory and general relativity using a common notion of real possibility and a common notion of empirical evidence; and, of course, general relativity now comes out as confirmed by the evidence (such as the advance of the perihelion of Mercury) in preference to the Newtonian theory. The remarkable observation, however, is that what is now seen to be preferentially confirmed by such evidence is exactly what distinguishes general relativity from the alternative formulation of Newtonian theory: namely, the (infinitesimally) Minkowskian metrical structure derived from special relativity, which here replaces the classical spatio-temporal structure based on absolute simultaneity. In other words, what is here preferentially confirmed by the evidence is exactly the special relativistic light principle. That which is characterized as a non-empirical constitutive principle, entirely beyond the reach of standard empirical testing, at one stage of scientific progress can be subject to precisely such testing at a later stage.[29] But this is just as it should be on a truly dynamical conception of the a priori.

---

and III.8). We can (uniquely) recover the traditional formulation given certain boundary conditions – such as a so-called "island universe" where a finite distribution of matter (e.g., the solar system) is effectively isolated from all other matter (in which case the traditional inertial trajectories are defined by the center of mass of the distribution). What we are doing here is essentially the same as Newton's use of the third law of motion to resolve the ambiguity created by Corollary VI (note 27 above): compare Stein (1977, section V). When the boundary conditions in question do not obtain, the traditional coordination (to a flat inertial structure) then fails; and such an eventuality would provide empirical motivations for preferring the new coordination, just as the empirical facts presupposed in the light principle motivate a preference for using it as the basis for a new coordination in special relativity. Given the classical boundary conditions, however, both the traditional coordination (based on the laws of motion) and the new coordination (based on the principle of equivalence) are equally well defined. Unlike the case of the light principle, therefore, the principle of equivalence does not by itself give us an empirical reason for rejecting or questioning the traditional classical coordination.

29. The light principle still counts as constitutive in general relativity, however, for we can still use it to define the infinitesimally Minkowskian metrical structure: see the reference cited in note 10 above. Indeed, in imagining carrying out the above empirical test of the light principle we are tacitly supposing the availability of an alternative empirical coordination for spatio-temporal metrical structure – such as one based (at least infinitesimally) on rigid bodies and physical clocks. This leads to subtle problems which cannot be discussed here: for some relevant discussion see Friedman (2001).

3

Rationality and Revolution

Lecture III outlined a conception of scientific rationality and applied it to the debates about this notion that have sprung up in the wake of Thomas Kuhn's theory of scientific revolutions. I argued that Kuhn's own attempt to defend a more traditional "universalistic" conception of scientific rationality against the threat of conceptual relativism others have seen as directly arising from his work does not succeed, and I then attempted to articulate a rather different defense based on the notion of *communicative rationality* derived from the philosopher Jürgen Habermas. In particular, I argued that Kuhn's defense fails by relying on what Habermas calls instrumental, as opposed to communicative rationality, and I suggested that the real problem arising from Kuhn's theory of scientific revolutions concerns this latter notion. Different constitutive frameworks or paradigms employ different – and even incommensurable or non-intertranslatable – standards of communicative rationality and precisely thereby raise the threat of conceptual relativism.

This point has been clarified, I hope, by the discussion at the beginning of section 2. Standards of communicative rationality are given by what I there call an empirical space of possibilities or space of reasons, in that agreement on the constitutive principles definitive of such an empirical space of possibilities (mathematical principles and coordinating principles) yields agreement on what can count as an empirical reason or justification for any given empirical possibility. A shared constitutive framework thereby facilitates shared mutually comprehensible rational argumentation, although, of course, it does not guarantee that agreement or consensus on the *results* of such argumentation will also necessarily be reached.[30] Thus, for example, those who share

---

30. The kind of consensus relevant to communicative rationality is therefore that necessary for mutual *understanding* – for (at least minimal) agreement on how to engage in rational deliberation with one another. But we do not thereby require agreement on everything, or even on very

enough of the new Einsteinian constitutive framework (defined, essentially,
by the Riemannian theory of manifolds and the principle of equivalence) can
agree that the advance of the perihelion of Mercury counts, in principle, as an
empirical test of Einstein's new theory of gravitation (the general relativistic
field equations) vis-à-vis the Newtonian theory of universal gravitation. But
there is still room for doubt, of course, about whether this test is truly decisive:
possible systematic errors and various competing hypotheses can still be
invoked in support of the Newtonian theory. Before the new constitutive
framework was actually available, by contrast, Einstein's new theory of gravitation was not even an empirical possibility, and no empirical test or
justification for it was therefore possible at all – not even one with a debatable
outcome.[31]

Viewing Einstein's successful calculation of the advance of the perihelion
of Mercury as a genuine empirical test of the general relativistic field equations thus implies a prior agreement on a constitutive framework that defines
the relevant notion of empirical reason or justification. Those (like a stub-
born, or perhaps simply uneducated, Newtonian physicist, for example) who
do not accept such a framework – those for whom Einstein's theory of gravitation is not even empirically possible – can at most accept the new theory as
a "black box" predictive device. For them, employing the new theory in this
purely predictive function may indeed be instrumentally rational in Habermas's sense: it may be a useful pragmatic "adaptation" for getting around in
the empirical environment. But the theory is not even a possible object of
communicative rationality – a possible candidate for mutual rational understanding and deliberation on the basis of evidence. My difficulty with Kuhn's
own defense of scientific rationality against the threat of conceptual relativism, then, is that he fails to distinguish these two very different notions. In
particular, when Kuhn speaks of an unproblematic increase of "puzzle-solving" success across scientific revolutions or paradigm shifts, he can only be
referring, in my terms, to the purely pragmatic or instrumental success to
which both the Newtonian and the Einsteinian physicist must necessarily

much. In this respect, shared standards of communicative rationality may be aptly compared
with shared principles of logic (although, as suggested above, they may also, in the case of empirical natural science, go well beyond the principles of formal logic and mathematics).
31. This is the crucial difference, once again, between standard cases of Duhemian empirical
underdetermination and genuine scientific revolutions. The distance between constitutive principles and empirical evidence is totally different from that arising from simple empirical
underdetermination, in that what is here in question is the very notion of empirical justification
or reason: compare notes 22 and 24 above.

agree.[32] Although it is entirely uncontroversial, as I have suggested, that this kind of purely instrumental success does indeed accrue to the later paradigm, it does not touch the real problem about scientific rationality raised by Kuhn's own historiographical work.

According to Kuhn himself successive paradigms in a scientific revolution are incommensurable or non-intertranslatable. There is an important sense, for example, in which practitioners of the earlier paradigm are incapable even of understanding the later paradigm. Kuhn compares successive paradigms, accordingly, to radically separate languages belonging to radically different cultural traditions.[33] And it is precisely in this connection that an obvious threat of conceptual relativism then arises: namely, what is rationally acceptable within one paradigm may not be so according to the standards (according to the "logic," as it were) of the other (as Einstein's theory of gravitation, in our example, cannot possibly be rationally acceptable within the context of a Newtonian constitutive framework). Pointing to the unproblematic purely instrumental success of the later paradigm entirely fails to respond to this threat, which rather concerns what we are now in a position to recognize as a quite fundamental problem within the domain of *communicative rationality*.[34]

Our problem, then, is to explain how a revolutionary transition from one scientific paradigm or constitutive framework to another can be communicatively rational, despite the fact that we are in this case faced with two essentially different and even incommensurable "logical spaces." Moreover,

---

32. Kuhn describes the goal of puzzle solving as establishing a "match between phenomena and belief." But precisely here lies a crucial ambiguity. It may refer, on the one hand, to a mere successful calculation or, on the other, to a genuine empirical test carried out against the background of an already accepted constitutive framework or paradigm.

33. This perspective on inter-paradigm incommensurability begins with section 5 of the Postscript to the second (1970) edition of *The Structure of Scientific Revolutions*, where Kuhn explicitly invokes the problem of radical translation as formulated in Quine (1960).

34. This close association of communicative rationality with scientific rationality and the theory of scientific revolutions is in fact entirely foreign to Habermas's own approach, for Habermas himself tends to associate scientific rationality with what he calls "cognitive-instrumental rationality" rather than communicative rationality. In particular, Habermas, like Kuhn, fails to distinguish between what I am calling a purely instrumental or pragmatic conception of scientific testing and a genuinely (communicatively) rational conception emphasizing prior agreement on a constitutive framework. In thus failing to acknowledge the profound contribution of mathematical and scientific knowledge to rational consensus and intersubjective communication, Habermas here places himself within the tradition of post-Kantian idealism that aims to defend the rightful claims of the *Geisteswissenschaften* against the "positivistic" presumptions of the *Naturwissenschaften*.

our commitment to a relativized yet still constitutive conception of the a priori only makes the problem more difficult. For this commitment implies that there is an important sense in which we must agree with Kuhn that successive paradigms, in a genuine scientific revolution, are actually non-intertranslatable: the later paradigm, from the point of the earlier paradigm, is not even a coherent possibility. How, therefore, can it ever be (communicatively) rational to accept the later constitutive framework? How, in particular, can there ever be empirical evidence that counts as an empirical reason, in our sense, in support of the later framework?

The discussion of inter-framework or inter-paradigm convergence in Lecture III is meant to be the first part of an answer to these questions. For, to the extent to which such convergence in fact obtains, the later space of possibilities can be seen as an *expansion* of the earlier one, so that the later constitutive framework contains the earlier as an (approximate) special case. Indeed, with respect to the purely mathematical part of our constitutive frameworks, we have the stronger result that the later principles contain the earlier principles quite exactly, in that the space of possibilities in mathematics continuously (and, as it were, monotonically) increases. The new calculus and analytic geometry developed in the seventeenth century, for example, retains all the theorems and principles of traditional Euclidean geometry intact, while adding further theorems and principles concerning a wide variety of new curves and figures (not constructible with straight-edge and compass) going far beyond traditional Euclidean geometry. Similarly, to take a somewhat different example, the new theory of Riemannian manifolds developed in the second half of the nineteenth century also retains Euclidean geometry as a perfectly exact special case (that of a three dimensional manifold of constant zero curvature), while adding a wide range of alternative possible geometries (of all dimensions and all possible curvatures, whether constant or variable) to the mix. Revolutionary transitions within pure mathematics, then, have the striking property of continuously (and, as it were, monotonically) preserving what I want to call *retrospective* communicative rationality: practitioners at a later stage are always in a position to understand and rationally to justify – at least in their own terms – all the results of earlier stages.

In mathematical physics, however, the situation is rather more complicated. For, in the first place, we are here dealing with approximate rather than exact containment, and, in the second place, we must here reckon with real (or empirical) as well as logico-mathematical possibility. As the space of real or empirical possibilities continuously (but not quite monotonically) expands, principles that count as constitutive at one stage may shift to the status of merely empirical laws at a later stage. And it is precisely because of this phenomenon, I believe, that revolutionary transitions within mathematical natural science give rise to a much more serious challenge of conceptual relativism.

Both of these points are clearly exhibited in our example of the transition from the classical Newtonian theory of gravitation to the general theory of relativity. In order rigorously to derive the classical Newtonian theory from general relativity as an approximate special case we need two distinguishable steps. In the first, we derive the alternative formulation of Newtonian gravity discussed at the end of section 2 from the space-time structure of general relativity by letting the velocity of light go to infinity, so that the light-cone structure present at each point in a general relativistic space-time manifold "collapses" into a Newtonian plane of absolute simultaneity at each point. The result is a Newtonian space-time structure of *four dimensional* variable curvature representing the effects of gravitation on non-flat space-time trajectories or geodesics (the "freely falling" trajectories), where, nonetheless, the *three dimensional* purely spatial structure on the planes of absolute simultaneity remains Euclidean or flat.[35] But we do not have the classical Newtonian laws of motion in this formulation, for the Newtonian law of inertia, in particular, has here been replaced by the principle of equivalence. In the second step, however, we recover the traditional formulation of Newtonian gravity from this modern (post-general-relativistic) alternative by considering a relatively isolated distribution of matter in the space-time of the modern formulation, where the center of mass of this distribution (which is perfectly well defined on any plane of absolute simultaneity) is then seen to describe a classical (four-dimensionally flat) inertial trajectory. In this special case the classical Newtonian laws of motion all hold, and the effects of gravitation reappear as classical accelerations produced by a classical inverse-square gravitational force.[36]

From the retrospective point of view of the new Einsteinian constitutive framework, therefore, we can, as suggested at the end of section 2, consider both the general relativistic field equations and the classical Newtonian law of gravitation as alternative empirical possibilities defined within a common empirical space of reasons. And, within this common constitutive framework

---

35. Intuitively, as the light-cone structure "collapses" into planes of absolute simultaneity, all purely spatial curvature is "squeezed out." For rigorous treatments of this limiting procedure see Ehlers (1981), Malament (1986).

36. Again, for further details on this second step see the discussion in note 28 above, together with the references cited there.

(given, essentially, by the Riemannian theory of manifolds and the principle of equivalence), it then appears that empirical evidence (such as the advance of the perihelion of Mercury) may clearly count in favor of the Einsteinian theory (again subject to the standard provisoes governing empirical underdetermination). This evidence clearly counts, in other words, in favor of the proposition that the invariant velocity of light (and, more generally, the invariant limiting velocity for the propagation of causal signals) is in fact finite. Analogously to the case of pure mathematics, then, the retrospective point of view of the new constitutive framework puts us in a position to understand the old framework (by way of reconstruction), to consider the new theory formulated within the new framework as better supported by the empirical evidence than the old theory (still from the point of view of the new framework), and to show (also from the point of the new framework) that the old theory nonetheless holds to a high degree of approximation in certain well defined special circumstances (relatively low velocities in comparison with light, relatively isolated distributions of matter).

The main disanology with pure mathematics, however, is not simply that the retrospective containment is here only approximate. The crucial point, rather, is that the later constitutive framework employs essentially different constitutive principles. In particular, the post-Einsteinian reconstruction of Newtonian gravitation theory is based on the principle of equivalence as its central coordinating principle, whereas the original formulation of Newtonian theory is of course based on the classical laws of motion. Further, whereas the classical laws of motion are constitutively a priori coordinating principles in the original formulation of Newtonian theory, in which the principle of equivalence appears as an additional, non-constitutive empirical fact, what is similarly constitutively a priori in the post-Einsteinian reconstruction is precisely the principle of equivalence, and the classical laws of motion now appear as mere empirical conditions holding in certain special empirical circumstances. (Strictly speaking, of course, the classical laws of motion are not even exactly valid empirically from a post-Einsteinian point of view.) What we have in fact recovered, therefore, is not the classical constitutive framework as such, but only an empirical counterpart to this classical framework formulated within an entirely different constitutive framework. The limiting procedure therefore fails to preserve constitutivity, even as an approximate special case.

We have here captured the sense, I believe, in which the Kuhnian claims of incommensurability and non-intertranslatability between successive frameworks in a scientific revolution are correct. The later framework is not translatable into the earlier framework, of course, simply because the concepts used in formulating the later framework have not yet come into existence. Before the invention of non-Euclidean geometries, for example, there is no way even to formulate the idea that space (or space-time) might have variable curvature; and this much, as the same example shows, is also true for revolutionary transitions in pure mathematics. In pure mathematics, however, there is a very clear sense in which an earlier conceptual framework (such as classical Euclidean geometry) is always translatable into a later one (such as the Riemannian theory of manifolds). In the case of coordinating principles in mathematical physics, however, the situation is quite different. To move to a new set of coordinating principles in a new constitutive framework (given by the principle of equivalence, for example) is precisely to abandon the old coordination based on the prior constitutive framework (given by the laws of motion, for example): what counted as coordinating principles in the old framework now hold only (and approximately) as empirical laws, and the old constitutive framework, for precisely this reason, cannot be recovered as such. By embedding the old constitutive framework within a new expanded space of possibilities it has, at the same time, entirely lost its constitutive (possibility defining) role.[37]

In these terms, therefore, the most fundamental problem raised by the Kuhnian account of scientific revolutions (and, in particular, by the idea of incommensurability) is to explain how it can be (communicatively) rational to move to a new constitutive framework, based on a radically new set of coordinating principles, despite the fact that this new framework, from the point of the old constitutive framework, is not even (empirically) possible. What rational motivations can there be – and how can it even be rationally intelligible – to make such a radical shift? Once *this* shift has been successfully negotiated, the rest of the story is relatively straightforward. For, as soon as the new empirical space of possibilities is in place (as soon as it is accepted as itself really possible), we can then use standard empirical testing, proceeding against the background of the new constitutive framework, differentially to confirm the empirical laws of the new framework relative to the (reconstructed) empirical laws of the old framework. (Retrospectively, that is, new

---

37. This captures the sense, in particular, in which there has indeed been a "meaning change" in the transition from the old framework to the new: even if the same terms and principles reappear in the new framework, they do not have the same meaning they had in the old, for they may no longer function as constitutive. And it is also worth noting that an analogous result would obtain in Carnap's theory of linguistic frameworks, where a change in status from analytic to synthetic would similarly involve a change of meaning (but compare also note 15 above).

empirical evidence can indeed confirm the new theory.) Yet this procedure, as we have seen, cannot possibly resolve our present problem. It cannot explain how it can be (communicatively) rational to move to the new space of possibilities in the first place, simply because rational empirical evidence, in this sense, presupposes that the new constitutive framework is already in place – and, as we have just seen, accepting the new framework implies, in general, a rejection of the old framework. Our problem, therefore, is not to explain why it is rational to accept Einstein's new theory of gravitation, for example, as correct. This question can be settled in a relatively straightforward fashion once the radically new coordination effected by Einstein's principle of equivalence is in place. Our problem is rather to explain how Einstein's new theory of gravitation becomes a rational or reasonable possibility in the first place – to explain, as it were, how it first became a live option. The empirical evidence counting unproblematically in favor of the *truth* of Einstein's new theory (such as the advance of the perihelion of Mercury) does not and cannot touch this latter question, which rather concerns, in our terms, the *real possibility* of the theory.

The first point to notice is that Kuhn's use of the analogy of radically disparate speech communities to elucidate the kind of incommensurability or non-intertranslatability at issue here is in fact entirely inappropriate. Succeeding conceptual frameworks in a scientific revolution are not aptly compared with radically separate languages or cultural traditions, for the new framework is self-consciously articulated against the background of the old framework. Indeed, this is precisely why it is so important, in the mathematical exact sciences, to recover the previous framework as a special case (whether exact or approximate).[38] Although the new framework may in fact be incommensurable with the old, in that, at the very least, it involves a genuine expansion of our space of intellectual possibilities, it nonetheless proceeds by an expansion – and thus a development – of that which was there before. In this respect, as I suggest in Lecture III, succeeding conceptual frameworks in a scientific revolution are more akin to different stages of development within a common linguistic or cultural tradition than they are to wholly disparate languages within radically separate cultural traditions.

---

38. This phenomenon appears to be distinctive of the mathematical exact sciences. In other areas of cultural and intellectual life it is typically more important, at moments of decisive cultural or intellectual revolution, to emphasize a radical break with the previous tradition rather than to conserve or to recover it. This is certainly true in the history of philosophy, for example. It is typical of the mathematical exact sciences, by contrast, that the most radical intellectual innovations are accompanied by equally fundamental efforts at recovery.

Now the relationship set up between succeeding conceptual frameworks by this type of inter-paradigm convergence is, as we have seen, a retrospective one. It is only from the point of view of the new framework that the earlier framework can be seen as a special case, so that, accordingly, our evolving space of intellectual possibilities can be truly seen as expanding (and thus as preserving, as far as possible, that which was there before). What we now need, as the second piece of our puzzle, is a *prospective* account of inter-para-digm (communicative) rationality suitable for explaining how, *from the point of view of the earlier framework*, there can still be some kind of (communica-tively) rational route leading to the later framework. The above observation about different stages of development within a common linguistic or cultural tradition has already put us on the right track, I believe; for what we are now in a position to add to the purely retrospective notion of convergence is the idea that the concepts and principles of the new constitutive framework do not only yield the concepts and principles of the old framework as a special case (whether exact or approximate), but they also develop out of, and as a natural continuation of, the old concepts and principles. In this way, the new constitutive framework is a quite deliberate modification or transformation of the old constitutive framework, developed against the backdrop of a com-mon set of problems, conceptualizations, and concerns. Despite the fact that practitioners of the new framework indeed speak a language incommensura-ble or non-intertranslatable with the old, they are nonetheless in a position rationally to appeal to practitioners of the older framework, and to do this, moreover, using empirical and conceptual resources that are already available at precisely this earlier stage.

Thus, Einstein's introduction of the radically new coordinating principles at the basis of the general theory of relativity (the light principle and the prin-ciple of equivalence) was self-consciously effected against the background of the quite different constitutive framework of classical mathematical physics – and it is entirely unintelligible without this background. Einstein's original introduction of the light principle in the special theory, for example, took for granted not only the immediately preceding work in the tradition of Lorentz-ian electrodynamics of moving bodies (Lorentz, Fitzgerald, Poincaré) but also the late nineteenth century work on the concept of inertial reference frame that definitively clarified the status of the question of absolute versus relative motion in classical Newtonian mechanics (see note 5 above). Ein-stein's introduction of the light principle was thus inextricably connected with what he calls the principle of relativity, and it thereby was seen to be a natural continuation, as it were, of the long tradition of reflection on the question of absolute versus relative motion extending back to the seventeenth century. Moreover, as we pointed out in section 2, Einstein here took an already well-established empirical fact (the empirical indistinguishability of different inertial frames by optical and electrodynamical means) and "elevated" it to the status of a convention or coordinating principle. What he saw, which no one did before, was that this already well-established empirical fact can indeed provide the basis for a radically new coordination of spatial and temporal structure and, at the same time, that it also calls into question the traditional classical coordination (based on the laws of motion and the possibility of instantaneous action at a distance), which everyone before Einstein had simply taken for granted (see note 23 above).

In the case of the principle of equivalence we have a similar story. That physics must employ a fundamental state of natural motion, such that deviations from this natural state are to be explained by external "forces," is an ancient idea going back to Aristotle. We have seen, moreover, how classical physics arrived at its own conception of natural motion by a continuous transformation, in the work of Galileo, of the preceding Aristotelian conception. Natural motion was now captured in the classical concept of inertia, and deviations from this natural state were explained by the precise mathematical concept of force finally articulated by Newton and expressed in his laws of motion.[39] In this way, as we have seen, the modern conception of natural motion became the basis for a new coordination of mathematical spatial and temporal structure to sensory experience. Already in classical physics, however, there was an intimate connection between gravitation and inertia (the equality of gravitational and inertial mass), and, despite the circumstance that it here appeared in the guise of an additional empirical fact (over and above the spatio-temporal coordination already effected by the laws of motion), it still played a crucially important role in the Newtonian theory of universal gravitation (see note 27 above). What Einstein did, once again, is "elevate" this already well-established empirical fact to the status of a fundamentally new coordination of spatio-temporal structure. Einstein thereby created a radically new conception of natural motion (where, in particular, gravitation no longer appears as a Newtonian external force) – but one which, nonetheless, naturally and continuously evolved from the preceding conception. Finally, and at the same time, Einstein also placed his radically new conception of the relationship between gravitation and inertia into the

---

39. For the intricate evolution of the Newtonian conception of force see Westfall (1971).

same long tradition of reflection on the question of absolute versus relative motion originally addressed in the special theory.[40]

In the case of both of his fundamental coordinating principles, then, Einstein directly appealed to already accepted empirical facts and to already established conceptual resources and problems. He thereby put practitioners of the earlier physics in a position both to understand his introduction of a radically new coordination and, with a little good will, to appreciate it and accept it as a genuine alternative. Moreover, whereas none of these considerations amounted to a rational *compulsion* to embrace the new paradigm, we have also seen that such a strong result is in no way required by the problem within the domain of communicative rationality on which we are focussing. To gain acceptance of the new framework merely as a rational (real) possibility – as a reasonable and responsible live option – is already more than half of the battle.

---

40. There are considerable conceptual pitfalls here, however. For, as is now well known, the general theory of relativity does not fully realize the "relativistic" ambitions by which Einstein was originally motivated; the general theory does not, in the relevant sense, amount to an "extension" of the special relativistic (and classical) principle of relativity. Nevertheless, the principle of equivalence does incorporate gravitational acceleration, in particular, into the traditional problematic of absolute versus relative motion in an especially striking and illuminating way. For discussion of this issue see, e.g., Friedman (1983, especially sections V.4 and V.5).

4

The Role of Philosophy

I have suggested that distinctively philosophical reflection plays a special and characteristic role in transitions between radically different conceptual frameworks during scientific revolutions. And I have suggested, accordingly, that we must distinguish three different levels of thought or discourse at work in such revolutions. Aside from the two already distinguished levels of properly empirical laws and constitutively a priori principles (both coordinating principles and mathematical principles), which belong, on this account, to the changing scientific paradigms themselves, we need also to distinguish a third or meta-scientific level where distinctively philosophical reflection takes place. Here we are concerned with what I want to call meta-paradigms or meta-frameworks, which play an indispensable role in mediating the transmission of (communicative) rationality across revolutionary paradigm shifts, despite the fact that they are incapable, by their very nature, of the same degree of (communicatively) rational consensus as first-level or scientific paradigms. In particular, when the concepts and principles of a later scientific paradigm develop through what I called, at the end of section 3, a "natural continuation" of the concepts and principles of an earlier one, reflection on the distinctively philosophical or meta-paradigmatic level helps us to define, during the revolutionary transition in question, what we now mean by a natural, reasonable, or responsible such continuation.

Thus, at the end of section 3, I urged that Einstein was able rationally to appeal to practitioners of the preceding paradigm in (classical) mathematical physics partly by placing his articulation of fundamentally new coordinating principles within the long tradition of reflection on the question of absolute versus relative motion going back to the seventeenth century. But this tradition of reflection is itself largely philosophical; and one way to bring this out in an especially vivid way is to keep in mind the large amount of unresolved intellectual disagreement surrounding it up to the present day. In the seventeenth century, for example, the various "natural philosophers" who all agreed on the first-level scientific paradigm of the mechanical philosophy – such as Descartes, Huygens, and Leibniz, for example – had sharply divergent things to say about absolute versus relative motion.[41] And it was precisely these disagreements, moreover, which, at least in part, fueled Newton's own articulation of both a radically different first-level scientific paradigm (based on the laws of motion and the possibility of instantaneous action at a distance) and a radically different answer to the question of absolute versus relative motion.[42] Then, in the eighteenth century, when Newton's first-level scientific paradigm was an unequivocal and uncontroversial triumph, problems and disagreements about absolute versus relative motion persisted nonetheless – as can be seen from the radical disagreements on this score between orthodox English Newtonians such as Samuel Clarke, on the one hand, and less orthodox continental Newtonians such as Leonhard Euler or Immanuel Kant, on the other.[43] In the nineteenth century we saw both a philosophical continuation of the debate in such writers as Ernst Mach and a scientific resolution, of sorts, in the development of the (classical) concept of inertial reference frame by Neumann, Thomson, and Lange (see again note 5 above); and it was precisely this nineteenth century background, as we just pointed out, that formed the immediate background to Einstein's own reintroduction of the problem. Nevertheless, although the new constitutive framework Einstein erected against this background was itself accepted as relatively clear and uncontroversial, at least with time, characteristically philosophical conceptual problems and debates about absolute versus relative motion still remained – even with respect to the status of these notions in Einstein's own theories (see again note 40 above).

Now, however, we have a new difficulty. If one of the characteristic differ-

---

41. What I am calling philosophical reflection need not, of course, be exclusively pursued by professional philosophers. Indeed, in the seventeenth century there was no clear distinction between professional philosophers and professional natural scientists. Nevertheless, we can distinguish, from our own perspective, between what we now call philosophical and natural scientific aspects of seventeenth century thought.

42. See Lecture II, note 53 above.

43. The Leibniz-Clarke Correspondence, which dominated intellectual discussion in "natural philosophy" for most of the first half of the eighteenth century, is a particularly clear indication of the distinctive status of this kind of reflection. This Correspondence provoked a lively and largely unresolved intellectual debate, despite the fact that virtually all parties to the debate accepted Newton's laws of motion and theory of gravitation as entirely unproblematic. And another clear indication of the distinctively philosophical status of the issues here is the way in which they became inextricably entangled with other characteristically philosophical problems, such as the nature of substance, God's relation to the natural world, freedom and necessity, and so on.

ences between philosophical and scientific reflection, as I am claiming, is that
the former necessarily fails to reach the (communicatively) rational consensus achieved by the latter, how can philosophical reflection possibly help in
mediating and (re)fashioning such rational consensus during scientific revolutions? How can a subject inevitably and permanently fraught with unresolved intellectual disagreements possibly help us to achieve a new rational
consensus in the case of a radically new scientific paradigm? The answer to
this difficulty is threefold. First, the consensus we require in the case of a radically new scientific paradigm is, as we have seen, relatively weak: we require
only that the new constitutive framework become a reasonable and responsible live option. Second, although we do not (and I believe should not) attain a
stable consensus on the *results* of distinctively philosophical debate, we do,
nonetheless, achieve a relatively stable consensus on what are the important
contributions to the debate and, accordingly, on what moves and arguments
must be taken seriously (see again Part One, notes 23 and 70). Third, characteristically philosophical reflection interacts with properly scientific reflection
in such a way that controversial and conceptually problematic philosophical
themes become productively intertwined with relatively uncontroversial and
unproblematic scientific accomplishments; as a result, philosophical reflection can facilitate interaction between different (relatively uncontroversial
and unproblematic) areas of scientific reflection, so as, in particular, to facilitate the introduction and communication of a new scientific paradigm at the
same time.

Once again, all three of these points are clearly exhibited in Einstein's
engagement with the problem of absolute versus relative motion. As we have
emphasized repeatedly, what Einstein needed, above all, was for his two new
coordinating principles (the light principle and the principle of equivalence)
to be taken seriously as a *possible* new constitutive framework for space, time,
and motion. He needed these principles to be taken seriously as a live alternative to the classical coordination based on the laws of motion; and it was to
this end, more specifically, that he appealed to the traditional debate about
absolute versus relative motion surrounding the classical constitutive framework. Moreover, at the distinctively philosophical level, where the traditional
debate continued, there was of course no widespread agreement about the
results of this debate, but there was, nonetheless, considerable agreement that
the key moves and arguments had been contributed by such thinkers as
Newton, Leibniz, Kant, Euler, and Mach. Finally, there had recently been a relatively uncontroversial and unproblematic scientific contribution to this
debate (with respect to the status of absolute versus relative motion within

classical Newtonian physics), namely, the late nineteenth century work on the concept of inertial reference frame referred to several times above. What Einstein did, in creating the new spatio-temporal coordination effected by the special theory of relativity, was to put this contribution into interaction with recently established empirical facts concerning the velocity of light in a striking and hitherto unexpected manner. Then, in creating the new spatio-temporal coordination effected by the general theory, Einstein, even more unexpectedly, put these two scientific accomplishments, together with the entire preceding philosophical debate on absolute versus relative motion, into interaction with a second already established empirical fact concerning the equality of gravitational and inertial mass. Since Einstein's introduction of a radically new conceptual framework was thus seriously engaged with both the established philosophical or meta-scientific tradition of reflection on absolute versus motion which had surrounded classical physics since its inception, and also with already established empirical and conceptual results at the scientific level, a classical physicist, on his own terms, had ample reasons seriously to consider Einstein's work. He did not, of course, need to adopt Einstein's new paradigm as correct, but he would have been irrational, unreasonable, and irresponsible (again on his own terms) to fail to consider it as a live alternative.

Yet these observations, as important as they are, are only one piece of an even more complex and interesting story of how philosophical or meta-scientific reflection contributed to the development, articulation, and promulgation of the general theory of relativity. Indeed, Einstein' engagement with the traditional problem of absolute versus relative motion does not by itself explain the most striking and significant revolutionary innovation of this theory – the introduction of non-Euclidean geometry into physics. To understand the truly astonishing way in which non-Euclidean geometry became intricately entangled with the problem of motion, we also need, as I suggested several times in the Lectures, to consider Einstein's engagement with the nineteenth century debate on the foundations of geometry initiated by Helmholtz and continued by Poincaré. Here we find an additional and extremely powerful source of philosophical mediation between general relativity and the preceding tradition in mathematical physics, without which, as we shall see, it is very hard to imagine how the use of non-Euclidean geometry in physics could have ever been envisioned as a real possibility.

Helmholtz, as I have suggested in the Lectures, began the debate in question by defending a version of geometrical empiricism. As I have also suggested, however, Helmholtz framed this empiricism within what he took to be a refinement of the Kantian view that space is a "necessary form of our spatial intuition." Indeed, it was precisely this generalized Kantian view of spatial intuition that set the stage for Helmholtz's own mathematical contribution to the foundations of geometry – a contribution which Helmholtz was originally motivated to pursue by his psycho-physiological researches into space perception.[44] Moreover, in developing this mathematical contribution, which we now know as the Helmholtz-Lie theorem, Helmholtz was directly inspired by Riemann's original (1854) work on the theory of manifolds. Helmholtz's goal, more specifically, was to derive Riemann's fundamental assumption, that the line-element or metric is Pythagorean or infinitesimally Euclidean, from what Helmholtz took to be the fundamental "facts" generating our perceptual intuition of space. Helmholtz's starting point was that our idea of space is in no way immediately given or "innate," but instead arises by a process of perceptual accommodation or learning based on our experience of bodily motion. Since our idea of space arises kinematically, as it were, from our experience of moving up to, away from, and around the objects that "occupy" space, the space thereby constructed must satisfy a condition of "free mobility" that permits arbitrary continuous motions of rigid bodies. And from this latter condition one can then derive the Pythagorean form of the line-element.[45] Since, however, the Riemannian metric thereby constructed has a group of isometries or rigid motions mapping any point onto any other, it must have constant curvature as well. So the scope of the Helmholtz-Lie theorem is much less general than the full Riemannian theory of metrical manifolds, which of course also includes manifolds of *variable* curvature.

The Helmholtz-Lie theorem fixes the geometry of space – and, according to Helmholtz, thereby expresses the "necessary form of our outer intuition" – as one of the three classical geometries of constant curvature: Euclidean, hyperbolic, or elliptic.[46] But how do we know which of the three classical

---

44. For Helmholtz's views on space perception see Hatfield (1990, chapter 5). For a discussion of Helmholtz's mathematical results in the context of his theory of space-perception see Richards (1977). See also Friedman (1997b). The following discussion draws heavily on Friedman (2001).

45. This result, as originally sketched by Helmholtz, was first rigorously proved by Sophus Lie within the latter's theory of continuous groups. For the work of Helmholtz and Lie see Torretti (1978, section 3.1). For a philosophically and mathematically sophisticated discussion of Helmholtz and Riemann see Stein (1977, section VI and VII – footnote 29, in particular, presents an up-to-date exposition of the mathematics of the Helmholtz-Lie theorem).

46. Helmholtz characterizes space as a "subjective *form* of intuition" in the sense of Kant, and as the "*necessary* form of our outer intuition," in his (1878) address on "The Facts in Perception" – see Hertz and Schlick (1921/77, p. 117/p. 124). Helmholtz viewed the condition of free mobility, in particular, as a necessary condition of the possibility of spatial measurement and thus of the application of geometry to experience. For discussion see the works cited in notes 44 and 45 above.

geometries actually holds? At this point, on Helmholtz's view, we investigate the actual behavior of rigid bodies (of rigid measuring rods, for example) as we move them around in accordance with the condition of free mobility. That physical space is Euclidean (which Helmholtz of course assumed) means that physical measurements carried out in this way are empirically found to satisfy the laws of this particular geometry to a very high degree of exactness. Thus Helmholtz's view was Kantian in so far as space indeed has a "necessary form" expressed in the condition of free mobility, but it was empiricist in so far as which of the three possible geometries of constant curvature actually holds is then determined by experience.

Now, as I have suggested, this Helmholtzian view of geometry set the stage, in turn, for the contrasting "conventionalist" conception articulated by Poincaré. Indeed, Poincaré developed his philosophical conception immediately against the background of the Helmholtz-Lie theorem, and in the context of his own mathematical work on group theory and models of hyperbolic geometry.[47] Following Helmholtz and Lie, Poincaré viewed geometry as the abstract study of the group of motions associated with our initially crude experience of bodily "displacements." So we know, according to the Helmholtz-Lie theorem, that the space thereby constructed has one and only one of the three classical geometries of constant curvature. Poincaré disagreed with Helmholtz, however, that we can empirically determine the particular geometry of space simply by observing the behavior of rigid bodies. No real physical bodies exactly satisfy the condition of geometrical rigidity, and, what is more important, knowledge of physical rigidity presupposes knowledge of the forces acting on the material constitutions of bodies. But how can one say anything about such forces without first having a geometry in place in which to describe them? We have no option, therefore, but to *stipulate* one of the three classical geometries of constant curvature, by convention, as a framework within which we can then do empirical physics.[48] Moreover, since

---

47. Poincaré discovered his well-known models of hyperbolic geometry in the context of his work on "Kleinian groups" in complex analysis – which he found, surprisingly, to include the isometries of hyperbolic geometry. This famous discovery is graphically reported in Poincaré (1908/13, Book I, chapter III). For a discussion of the Poincaré models see Torretti (1978, section 2.3.7).

48. The argument here, more specifically, is that establishing mathematical force-laws (underlying the physical notion of rigidity) presupposes that we already have a geometry in place in order to make spatial measurements; so we must *first* choose a particular geometry and then subsequently investigate physical forces. For a detailed analysis of Poincaré's argument along these lines see Friedman (1999, chapter 4). For further discussion of the relationship between Helmholtz and Poincaré – against the background of the original Kantian conception of geometry – see Friedman (2000a).

Euclidean geometry is mathematically the simplest, Poincaré had no doubt at
all that this particular stipulation would always be preferred.

I observed in the Lectures that Einstein was intensively reading Poincaré's
Science and Hypothesis as he was creating the special theory of relativity in
1905, and I suggested, accordingly, that Poincaré's conventionalism thereby
played a significant role in philosophically motivating this theory (see Part
One, note 27, together with the paragraph to which note 76 is appended).
More specifically, whereas Poincaré had argued, against both Kant and Helmholtz, that the particular geometry of space is not dictated by either reason or
experience, but rather requires a fundamental decision or convention of our
own, Einstein now argued similarly that simultaneity between distant events
is not dictated by either reason or experience, but requires a new fundamental
definition based on the behavior of light.[49] Moreover, as we have seen, Einstein
proceeded here, in perfect conformity with Poincaré's underlying philosophy
in *Science and Hypothesis*, by "elevating" an already established empirical fact
into the radically new status of what Poincaré calls a "definition in disguise" –
namely, what we call a coordinating principle (see again note 22 above).

However, as Einstein tells us in his famous 1921 paper on "Geometry and
Experience," he needed to reject Poincaré's geometrical conventionalism in
order to arrive at the general theory of relativity. In particular, Einstein here
adopts a Helmholtzian conception of (applied or physical) geometry as a
straightforward empirical theory of the actual physical behavior of "practically rigid bodies," and he claims, in a striking passage, that "without [this
conception] I would have found it impossible to establish the [general] theory of relativity." Immediately thereafter, in the same passage, Einstein
considers Poincaré's geometrical conventionalism – apparently as the only
real alternative to his own view – and points out (perfectly correctly) that "if
one [following Poincaré] rejects the relation between the practically rigid
body and geometry one will in fact not easily free oneself from the convention
according to which Euclidean geometry is to be held fast as the simplest." Einstein concedes that "[s]ub specie aeterni Poincaré, in my opinion is correct,"
for "practically rigid bodies" are in fact unsuitable to play the role of "irreducible elements in the conceptual framework of physics." Nevertheless, Einstein
suggests, they must *provisionally* "still be called upon as independent elements in the present stage of theoretical physics" – when, in particular, we are

49. Indeed, Poincaré had himself argued that distant simultaneity requires a convention or definition (also involving the velocity of light) in his 1898 article on "The Measurement of Time,"
reprinted as Poincaré (1905/13, chapter II). For discussion of the relationship between Poincaré's
views on simultaneity and Einstein's work see Miller (1981, chapter 4).

still very far from an adequate micro-theory of the structure of matter. And, where such rigid bodies must "still be called upon as independent elements," it is clear, is precisely in the foundations of the general theory.[50]

Here, as Einstein explains in the same passage, he has in mind the following line of thought. According to the principle of equivalence gravitation and inertia are essentially the same phenomenon. So, in particular, we can model gravitational fields by "inertial fields" (involving centrifugal and Coriolis forces, for example) arising in non-inertial frames of reference. If we consider a uniformly rotating frame of reference in the context of special relativity, however, we find that the Lorentz contraction differentially affects measuring rods laid off along concentric circles around the origin in the plane of rotation (due to the variation in tangential linear velocity at different distances along a radius), whereas no Lorentz contraction is experienced by rods laid off along a radius. Therefore, the geometry in a rotating system will be found to be non-Euclidean (the ratio of the circumference to the diameter of concentric circles around the origin in the plane of rotation will differ from π and depend on the circular radius).

The importance of this line of thought for Einstein is evident in virtually all of his expositions of the general theory, where it is always used as the primary motivation for introducing non-Euclidean geometry into the theory of gravitation.[51] Moreover, as John Stachel has shown, this particular thought experiment in fact constituted the crucial breakthrough to what we now know as the mathematical and conceptual framework of general relativity. For, generalizing from this example, Einstein quickly saw that what he really needed for a relativistic theory of gravitation is a *four dimensional* version of non-Euclidean geometry (comprising both space and time). He quickly saw that a variably curved generalization of the flat Minkowski metric of special relativity should serve as the representative of the gravitational field, and, turning to the mathematician Marcel Grossmann for help, he then discovered the Riemannian theory of manifolds. Einstein's repeated appeal to the example of the uniformly rotating frame of reference in his official expositions of the theory therefore appears to reflect the actual historical process of discovery very accurately, and to explain, in particular, how the idea of a variably curved four dimensional space-time geometry was actually discovered in the first place.[52]

---

50. See Einstein (1921/23, pp. 6–8/ pp. 33–6).

51. See, e.g., Einstein (1916/23, pp. 774–6/ pp. 115–7), (1917/20, §§ 23–8), (1922, pp. 59–61).

52. See Stachel (1980/89). It is especially striking, in this connection, that Einstein had at first dismissed Minkowki's (1908) geometrical interpretation of special relativity as a mere mathematical curiosity. For Einstein, unlike Minkowski, the idea of a *four dimensional* space-time geometry

The way in which Einstein came to use the principle of equivalence as a fundamental coordinating principle for the application of four dimensional (semi-)Riemannian geometry in mathematical physics is thus extraordinarily subtle and complex. When Einstein first formulated the principle of equivalence he had no conception of four dimensional (semi-)Riemannian geometry at all, and he was working, instead, within what we now conceive as the flat (or semi-Euclidean) four dimensional geometry of special relativity. Moreover, at this time, he did not even understand the special theory of relativity in four dimensional (Minkowskian) terms, but rather persisted in the traditional three-plus-one dimensional description in terms of (three dimensional) moving frames of reference.[53] In first applying the principle of equivalence in the context of this (three-plus-one dimensional) understanding of the special theory, he discovered a non-Euclidean *spatial* geometry in the case of the uniformly rotating frame of reference – which he only subsequently generalized thereafter to a non-Euclidean *space-time* geometry. It was only at this point, then, that the principle of equivalence actually functioned as we conceive it today: as the fundamental interpretive principle for the variably curved (semi-)Riemannian geometry described by Einstein's field equations, coordinating the mathematical notion of (semi-)Riemannian geodesic to the "freely falling" trajectories in a gravitational field. And it was only by this circuitous route, therefore, that the so far merely mathematical possibility of a non-Euclidean space-time structure first became a real, physical, or empirical possibility.[54]

Nevertheless, Einstein's engagement with nineteenth century philosophical or meta-scientific reflection on the foundations of geometry exhibits the three features noted above in connection with his parallel engagement with the problem of absolute versus relative motion. First, the crucial point he

---

and the idea of a *variably curved* four dimensional space-time geometry therefore arose together, in that it was only after the creation of the general theory that Einstein himself adopted a four dimensional interpretation of the special theory.

53. Compare again note 52 above. See especially Norton (1985/89) for a very clear and illuminating discussion of Einstein's use of the principle of equivalence at the time.

54. The only person, at the time, who could have hit upon the application of four dimensional (semi-)Riemannian geometry more directly was Hermann Minkowski, but Minkowski himself, as further discussed in Friedman (2001), had no idea at all of using the principle of equivalence to construct a relativistic theory of gravitation. Nor can Riemann's intriguing remarks speculating that the metric of physical space might be empirically determined by "binding forces" acting on the underlying manifold be taken as an anticipation of general relativity – as suggested by Weyl's well-known sixth and final *Erläuterung* to the final section of Riemann (1919). For Riemann had no inkling of the crucial application of the principle of equivalence, and his remarks thus remain *mere* speculations within the realm of mathematical (but not yet empirical or real) possibility.

needed to establish was that the radically new space-time structure in question
is indeed a real or empirical *possibility*, so that, in particular, the new theory of
gravitation could be seen as a live alternative to classical Newtonian theory.
Second, although there was of course no consensus on the results of the philosophical debate between geometrical empiricism and conventionalism (and
there is still not even to this day), there was nonetheless widespread agreement
that the key moves and arguments in this debate had been contributed by
Helmholtz and Poincaré. Third, this philosophical debate was also intimately
intertwined with recent, relatively uncontroversial results within the sciences,
namely, with mathematical work within the tradition of the Riemannian theory of manifolds and group theory that culminated in what we now know as
the Helmholtz-Lie theorem (where, as we have seen, Helmholtz and Poincaré
were themselves central contributors). Thus, anyone reasonably and responsibly concerned with the status of physical geometry – and the question, in
particular, of the application of non-Euclidean geometry in physics – had no
choice but to consider the philosophical reflections of Helmholtz and
Poincaré with the utmost seriousness. And, since Einstein's introduction of his
revolutionary new framework for space, time, and motion was also intensively
(and responsibly) engaged with precisely these philosophical reflections
(which it strikingly and unexpectedly integrated, moreover, with the traditional problem of motion), one had very little choice, at this point, but to
consider Einstein's new theory with the very same utmost seriousness.[55]

Finally, Einstein's engagement with nineteenth century philosophical or
meta-scientific reflection on the foundations of geometry illustrates an additional and very important feature of the role of such reflection during truly
revolutionary scientific transitions. In particular, when we move from one
scientific conceptual framework to a radically different one, there is necessar-

---

55. The reader may still wonder why Einstein had to side with Helmholtz rather than Poincaré at
this stage. This is a rather complicated matter that is further discussed in Friedman (2001). But
the point, briefly, is that Einstein, as he suggests, needed a "naïve" perspective on the relationship
between geometrical structure and "practically rigid bodies" according to which all questions of
micro-physics could at least provisionally be ignored. Just as, in the special theory of relativity,
Einstein takes the Lorentz contraction as a direct indication of fundamental *kinematical* structure, independently of all (dynamical) questions about the micro-physical forces actually
responsible for physical rigidity, here, in the example of the uniformly rotating reference frame,
Einstein similarly takes the Lorentz contraction as a direct indication of fundamental *geometrical*
structure. It is also worth noting that Einstein later came to revise this "naïve" perspective on the
relationship between geometrical structure and micro-physics in connection with his own work
on unified field theory. This revision is reflected in his well-known later discussion of geometrical
"conventionalism" in response to Hans Reichenbach in Schilpp (1949, pp. 676–9).

ily an intermediate stage in which we are still in the process of (continuously) transforming the earlier framework but have not yet clearly articulated the later one. There necessarily comes a point, as it were, when we are operating within neither the one nor the other and are, in fact, caught in a deeply problematic (but nevertheless intensely fruitful) state of inter-paradigmatic conceptual limbo. This is illustrated, in the present case, not only by the circumstance that Einstein first applied the principle of equivalence to what we now conceive as the flat geometrical structure of Minkowski space-time (where, more specifically, there is as yet no four dimensional space-time curvature), but, even more strikingly, by the fact that the preceding philosophical debate on the foundations of geometry was framed by the Helmholtz-Lie theorem and was thereby limited to spaces of constant curvature. Since the whole point of the general theory of relativity, in the end, is to describe gravitation by a four dimensional manifold of *variable* curvature, there is an important sense in which the final articulation of the general theory rendered the entire preceding debate irrelevant. Nevertheless, as we have seen, Einstein's final articulation and elaboration of this theory was essentially, and rationally, mediated by precisely this philosophical debate – without which, as we have also seen, it is indeed hard to imagine how the application of non-Euclidean geometry in physics could have ever become a real possibility and thus a genuinely live alternative.[56]

---

56. For further discussion of both of these conceptually problematic features of Einstein's initial application of the principle of equivalence see Friedman (2001), which here draws on Norton (1985/89). In particular, there is a way, even from our contemporary post-general-relativistic point of view, in which both Einstein's application of the principle of equivalence to a flat Minkowski space-time and his discovery of a non-Euclidean geometry therein in the context of the Helmholtz-Poincaré debate make perfect sense. For, if we contrast what we now know as the space-time of general relativity with that of the (variably curved) alternative formulations of Newtonian gravitation theory discussed above, we see that the crucial difference between them is precisely the infinitesimally Minkowskian structure of the former space-time structure. It is applying the principle of equivalence to the inertial structure of Minkowski space-time, therefore, that results in the characteristic empirical content of Einstein's new theory (such as the general relativistic derivation of the advance of the perihelion of Mercury). Moreover, it is only within the context of a relativistic, infinitesimally Minkowskian space-time structure that four dimensional space-time curvature is necessarily associated with three dimensional *spatial* curvature (note 35 above). Finally, as Norton shows, the three dimensional spatial geometry Einstein discovers in the example of the uniformly rotating reference frame is itself perfectly well defined, on the basis of infinitesimal approximations by an infinite number of differing "tangent" inertial frames – and thus, as Einstein says, on the basis of *infinitesimally* "rigid" measuring rods. (As I pointed out in note 40 above, a parallel situation of deeply problematic yet fruitful conceptual limbo also obtains in the case of Einstein's appeal to the traditional debate surrounding absolute versus relative motion.)

5

Other Problems, Other Sciences

I have here attempted to develop a modified version of a Kantian scientific epistemology capable of doing justice to the revolutionary changes within the sciences that have led to the most important philosophical challenges to the original Kantian conception. I have paid particular attention to the challenge posed by Quinean epistemological holism, which takes the post-Kantian revolutions in geometry and mechanics (among other things) as grounds for a naturalized empiricism in which there is nothing left of the a priori at all; and to the challenge posed by Kuhn's theory of scientific revolutions, which has led, in recent years, to a new style of historical and conceptual relativism based on the incommensurability or non-intertranslatability of succeeding revolutionary paradigms. Against Quinean holism I have argued for the importance of relativized yet still constitutive a priori principles in understanding the (evolving) conceptual frameworks at the foundations of modern mathematical physics; and against post-Kuhnian conceptual relativism I have argued that, despite the important kernel of truth in Kuhn's doctrine of incommensurability or non-intertranslatability, there is still considerably more rational continuity in the revolutionary transitions in question than either Kuhn or the post-Kuhnian conceptual relativists allow.

It is important to note, in this last connection, that what I have here offered in response to Kuhn's theory of scientific revolutions is a defense of scientific rationality, not a defense of "scientific realism." For me, the main problem posed by post-Kantian scientific developments is precisely a challenge to the idea of universal (trans-historical) scientific rationality – to the Enlightenment ideal of a fixed rationality, paradigmatically expressed in modern mathematical science, which can serve as the basis for rational argumentation and communication between all human beings. I have attempted to show that, although there are indeed no fixed principles of rationality of the kind Kant originally envisaged (based on the constitutive principles of specifically Newtonian mathematical physics), there is nonetheless sufficient (communicatively) rational continuity during revolutionary changes of constitutive principles so that a dynamical version of the Enlightenment ideal can still be maintained. The mathematical exact sciences still serve as the very best exemplars we have of universal communicative rationality in spite of, and even because of, their profoundly revolutionary character.

Just as Kant's original defense of scientific rationality did not proceed on the basis of what he called "transcendental realism," the idea that our human system of representations somehow corresponds to an otherwise entirely independent realm of "things in themselves," the present defense does not proceed on the basis of "scientific realism." In particular, although I have argued for an essential element of convergence in the historical evolution of successive constitutive frameworks, this is explicitly not convergence to an entirely independent "reality" (however conceived) but rather convergence *within* the evolving sequence of constitutive frameworks itself.[57] Indeed, at the end of Lecture III I suggested that the present conception of scientific rationality does not even require that there be a *uniquely* correct sequence of convergent successor theories – something that would certainly be required by any version of "scientific realism." However, just as in the case of Kant's original conception, we do retain an element of "internal" or what Kant called "empirical realism." For, once a given constitutive framework is already in place, there is a perfectly precise sense in which we can then speak of a "matching" or "correspondence" between a theory formulated within that framework and the empirical or phenomenal world: namely, when we have precise rational evidence, in the sense of section 2 above, for a properly empirical law in the context of an empirical test of that law. Within a given space of empirical possibilities, in other words, it makes perfectly precise sense to speak of empirical truth, and theories of mathematical physics continue to provide paradigmatic examples of such truth. In order to defend scientific rationality against the threat of post-Kuhnian relativism, however, there is no need to contemplate the "empirical truth" (whatever this might mean) of the changing constitutive frameworks themselves.[58]

---

57. Kuhn sometimes rejects the very idea of convergence on the basis of a prior rejection of "scientific realism." See especially (1962/70, pp. 206–7), where Kuhn rejects all talk of convergence over time on the grounds that "[t]here is, I think, no theory-independent way to reconstruct phrases like 'really there'; the notion of a match between the ontology of a theory and its 'real' counterpart in nature now seems to me illusive in principle." Compare also Part One, notes 72 and 73 above.

58. Here I am indebted to discussions at Stanford with Peter Godfrey-Smith and, especially, with Graciela De Pierris. There is an important difference at precisely this point between the conver-

In reacting to these post-Kantian problems and challenges I have concentrated on the development of our most fundamental mathematical-physical theories of space, time, and motion: on Newtonian mechanics and gravitation theory (against the background of the Aristotelian theory of space, time, and motion) and on the historical development of the revolutionary new geometrical and mechanical theories created by Albert Einstein – the special and the general theories of relativity. Aside from the circumstance that these are the theories with which I am myself most familiar (both historically and conceptually), there are several reasons for focussing on them. First, it was of course the Newtonian theory of space, time, and motion that provided Kant with his paradigm of scientific rationality and provided the basis, accordingly, for his original philosophical conception. Second, it has been precisely the revolutionary changes within our theories of space, time, and motion wrought by Einstein that have provided the primary motivations for post-Kantian attempts to reconsider the nature of scientific rationality – both in the case of the logical empiricists and in the case of the Kuhnian theory of scientific revolutions developed against the background of their thought. Third, it is within our modern mathematical theories of space, time, and motion, I believe, that we find the clearest examples of constitutively a priori coordinating principles and thus the clearest examples of genuine scientific revolutions: scientific transitions involving incommensurable or non-intertranslatable conceptual frameworks.

Nevertheless, the reader will naturally wonder how the present philosophical account bears on other cases of scientific revolutions standardly so-called; for, if it does not bear on them all, it would appear to be of very limited interest in our present post-Kuhnian environment. And the three cases that come most naturally to mind here are the quantum mechanical revolution (which has, in fact, been the most dominant influence by far on twentieth century mathematical physics), the chemical revolution of the late eighteenth century associated with the work of Antoine Lavoisier, and the Darwinian revolution in biology. Although, as I have said, the present philosophical framework has been fashioned in light of the development of our modern theories of space,

---

gence involved on the present account and the Peircean convergence alluded to in Lecture III (see Part One, notes 78 and 83). As on the present account, Peircean convergence is *inter*-theoretic, and thus does not involve the idea of convergence to an independently existing reality. Indeed, for Peirce "reality" is simply to be *defined* as whatever our best scientific theorizing eventually converges to. For precisely this reason, however, Peircean convergence is necessarily unique. Peircean convergence belongs to an account of (empirical) scientific truth, whereas ours belongs to an account of scientific rationality (and, in particular, the a priori constitutive principles definitive of such rationality).

time, and motion, and, as I should also admit, I am by no means prepared to give them here the attention they deserve, I shall attempt, in what follows, to say just a few words about these other three revolutionary transitions – which, if nothing else, may possibly serve to stimulate further philosophical and historical work.

The Quantum Mechanical Revolution. A consideration of this case is perhaps most pressing of all for our present account, for we here have a fundamental theory of modern mathematical physics – and certainly the most empirically successful one so far – where a precise mathematical framework or structure (a non-commutative algebra of Hermitean operators on a Hilbert space) is used to represent a corresponding system of physical entities (the physical quantities or "observables" characterizing atomic and subatomic objects). And it is here, in fact, that the matter (and energy) that "fills" our otherwise empty space-times (described by our best current theories of space, time, and motion) is theoretically characterized. Nevertheless, as is well known, despite the clarity and precision of the relevant mathematical structures, as well as the incredible empirical success that has accompanied their physical application, there remain deep conceptual problems surrounding the foundations of this theory – which, in recent years, have been increasingly seen to arise from difficulties in unifying or synthesizing the conceptual framework of quantum mechanics with those of our best contemporary space-time theories (both the special and the general theories of relativity). In this connection, I believe, although the present philosophical account is by no means in a position to solve these fundamental problems, it may shed light on their underlying sources and point us in promising directions.

The first point I want to make is that we have not seen the kind of fruitful interaction between scientific and philosophical or meta-scientific ideas in the case of quantum mechanics that we have seen in the case of the other great revolutions in modern mathematical physics. In the case of the quantum mechanical revolution, in particular, we have not seen *timely* interventions from the philosophical or meta-scientific realm, suitable for rationally bridging the gap between pre-revolutionary and post-revolutionary conceptual landscapes. To be sure, a wide variety of philosophical ideas have in fact been brought to bear in discussions of the foundations of quantum mechanics: such as Werner Heisenberg's appeal to the Aristotelian notion of potentiality, the appeal to Kantian views about the necessary structure of our human sensory and conceptual apparatus often associated with Niels Bohr's "Copenhagen" interpretation, Eugene Wigner's appeal to mind-body dualism, and the appeal to the concept of alternative possible worlds standardly associated with Hugh Everett's "relative state" interpretation.[59] However, unlike the case of Einstein's appeal to preceding nineteenth century philosophical debates concerning the problem of absolute versus relative motion and the foundations of geometry, for example, these appeals to philosophical ideas in the interpretation of quantum mechanics have not invoked ongoing traditions of meta-scientific reflection that are intimately intertwined with the pre-revolutionary conceptual situation actually addressed by the creation of the new paradigm. In this sense, appeals to Aristotelian potentialities, Kantian necessary cognitive structures, Cartesian mind-body dualism, or Leibnizean possible worlds come entirely out of the blue, as it were, and are not related in a timely fashion to the radically new empirical and conceptual problematic that has in fact given rise to quantum mechanics.

The unfortunate fact, from our present point of view, is that quantum mechanics simply has not been integrated with an ongoing meta-scientific tradition at the philosophical level at all. In the case of relativity theory, by contrast, not only was Einstein's initial creation of the theory essentially mediated, as we have seen, by a preceding philosophical tradition associated with Mach, Helmholtz, and Poincaré (among others), but Einstein's work was also intimately involved, in turn, with the creation of a new tradition of meta-scientific reflection, the logical empiricist tradition, which took Einstein's new theories as the basis for a radical reformulation of scientific philosophy. So far, however, quantum theory (despite the important clarifications contributed by professional philosophers of physics) has had no similar impact on the ongoing practice of philosophy as a discipline – even within those parts of the discipline focussed on epistemology and the philosophy of science. Moreover, although it may well be the case, of course, that this situation is simply one more inevitable effect of the increasing specialization of knowledge, it is also entirely possible that it rather represents a striking confirmation of the present account – in that it is precisely because of a lack of timely interaction with an ongoing tradition of philosophical reflection that the present difficulties in the conceptual foundations of quantum mechanics (especially with respect to a possible integration with our best current theories of space and time) have proved so profoundly intractable.

---

59. For the appeal to the Aristotelian notion of potentiality see Heisenberg (1958). Bohr's insistence on "the necessity of classical concepts" has frequently been attributed to the influence of Kantian and neo-Kantian ideas, but it is not entirely clear that Bohr himself thought of it in this way. For the appeal to mind-body dualism see Wigner (1961/67). Although Everett himself did not explicitly invoke the concept of alternative possible worlds, his viewpoint quickly became identified as the "many-worlds" interpretation: see DeWitt and Graham (1973).

My second point is at once more constructive and more speculative. In looking at the conceptual development of quantum mechanics from the point of view of the present account, one naturally looks for fundamental constitutive principles (especially coordinating principles) analogous to the Newtonian laws of motion and the Einsteinian principle of equivalence. And here perhaps the most plausible candidate is the so-called correspondence principle, which Bohr invoked as absolutely central to the theory from his early work on atomic spectra through his introduction of the idea of complementarity. The correspondence principle was explicitly intended by Bohr as the fundamental bridge between empirical phenomena, on the one side, and the new theoretical conception of the internal structure of the atom, on the other; and it performed this essential coordinating function by relating experimental phenomena to limited applications of *classical* concepts within the new evolving theory of atomic structure. For Bohr, moreover, it appears that it was precisely the correspondence principle which then led to the idea of complementarity. For, in applying this principle to different experimental phenomena, different and in some sense incompatible applications of classical concepts emerged: applying the correspondence principle to experimental results on atomic spectra, for example, resulted in the application of the concept of orbital periodic frequencies within the atom (a classical "space-time" description), whereas applying it to atomic scattering experiments, by contrast, led to the application of the concept of precise momentum and energy exchanges (a classical "causal" description).[60] My suggestion, then, is that a better understanding of the correspondence principle, both historically and conceptually, might shed considerable light on the present difficulties in the foundations of the theory by showing how empirical phenomena are (rationally) related to the new non-commutative mathematical structures in a way that is both systematic and principled.

My third point is even more speculative. I here want to observe that there has been one intervention in the foundations of quantum mechanics from the philosophical or meta-scientific realm which may have a better claim to timeliness in the present sense. In the 1930s the mathematician John von Neumann, who had worked intensively on both the current situation in the foundations of logic and mathematics and on the mathematical foundations of quantum mechanics, suggested that a revision of classical logic, wherein

---

60. Here I am indebted to discussions of the correspondence principle with Scott Tanona, who is writing a dissertation on this topic at Indiana University. (Tanona's work is still preliminary, however, and he is not to be held responsible for anything I say here.)

the distributive law for disjunction and conjunction is no longer universally valid, might be the best way for rigorously comprehending the radically new conceptual situation Bohr had attempted to address through the idea of complementarity. From this point of view, the logical space governing quantum mechanical quantities or properties is a non-distributive or non-Boolean algebraic structure, consisting of a system of distributive or Boolean sub-structures that cannot simultaneously be realized or embedded within a single comprehensive Boolean structure. And this yields a precise algebraic interpretation of the idea of complementarity (a non-orthodox one, of course), in so far as quantum-mechanical quantities or properties are indeed individually classical (within a particular Boolean sub-structure), and the underlying source of quantum-mechanical "weirdness" is simply that they do not fit together within a single classical (or Boolean) logical structure.[61] This interpretive suggestion, to my mind, represents the right *kind* of intervention from the philosophical or meta-scientific level: it engages with other plausibly relevant work in the foundations of the sciences (in this case plausibly relevant work in the contemporaneous foundations of mathematics), and, in more-or-less deliberate interaction with some of the best current philosophical reflection on these matters, it indicates a way in which the idea of a relativized and dynamical a priori can even extend to fundamental principles of logic.[62] It remains to be seen, however, whether von Neumann's interpretive suggestion can yield a full and satisfying resolution of the deep conceptual difficulties still afflicting the theory.[63]

---

61. This suggestion is first developed in Birkhoff and von Neumann (1936). Von Neumann put it forward as an alternative interpretation to (Bohrian) complementarity in a public discussion with Bohr in 1938: see his comments on Bohr (1939, pp. 30–9). The reader will recall from Lecture II that Quine (1951/53) appeals to the possibility of a quantum mechanical revision of classical logic; and Quine (1970, pp. 85–6) indicates that it is precisely von Neumann's work that is at issue here (although in both places Quine mistakenly locates the deviation from classical logic in the law of excluded middle).

62. Von Neumann participated in a well-known philosophical debate on the foundations of mathematics in 1930, where he represented the position of formalism, Arend Heyting the position of intuitionism, and Carnap the position of logicism; a translation of this debate can be found in Benacerraf and Putnam (1964/83). My own suggestion is that quantum logic should be understood within the Carnapian theory of linguistic frameworks, where it appears as an empirically motivated revision of fundamental constitutive principles but not, *pace* Quine (see note 61 above), as a testable empirical theory.

63. It does appear that the quantum logical point of view sheds considerable light on the problem of unifying quantum mechanics with the special theory of relativity highlighted by the discovery of Bell's inequality. In the EPR-Bohm experiment, for example, the singlet state appears as the conjunction of three disjunctions, each of which asserts a perfect anti-correlation for one of the three relevant (paired) directions of spin. The problem Bell's inequality poses for an interpreta-

The Chemical Revolution. Lavoisier's creation of a new system of chemistry, based on a radically new conception of combustion and calcination opposing the then dominant phlogiston theory, does not fit neatly into the present account of revolutionary paradigm shifts. The main conceptual innovation did not involve a radically new application of a (possibly novel) mathematical structure, and so constitutive and coordinating principles of the kind we have focussed on did not figure essentially in the new conceptual framework. Nevertheless, Lavoisier's revolution did involve a new application of what is plausibly an a priori constitutive principle of classical mathematical physics, namely, the principle of the exact conservation of the quantity of matter; and it thereby introduced rigorous experimental methods, based on a precise use of the balance, into the exact tracking of all products and agents (including especially gaseous ones) involved in chemical reactions. What I want to emphasize here, however, are two important ways in which the chemical revolution of the late eighteenth century is also interestingly connected with some of the central elements of our present story.

In the first place, the chemical revolution played a very important role in Kant's attempt to comprehend the radical scientific changes taking place around the turn of the century within the already established framework of his overall philosophical system – including both the *Critique of Pure Reason* of 1781 (revised in 1787) and the *Critique of Judgement* of 1790. In the period of the two editions of the *Critique of Pure Reason* Kant took phlogistic chemistry as his model for that discipline and, at the same time, denied full scientific status to chemistry. But in the following years Kant gradually assimilated the new discoveries associated with the work of Lavoisier, came to view chemistry as finally established as a genuinely scientific discipline, and strove to modify his preceding model of science accordingly. In fact, in the years from 1796 almost until his death in 1804 Kant was working on sketches and drafts of a new philosophical work, entitled *Transition from the Metaphysical Foundations of Natural Science to Physics*, where, among other things, he was very intensively occupied with the totality of scientific problems and discoveries

---

tion of this statement within classical probability theory is then seen to arise from an illegitimate application of the distributive law. Moreover, the (classically) problematic probabilistic correlations predicted by quantum theory also emerge as direct logical consequences of the singlet state: in a quantum logical probability space there is (by Gleason's theorem) one and only one probability measure consistent with that state (considered as a proposition), namely, the quantum mechanical one. In this sense, the (classically) problematic correlations are fully "explained" by an underlying "common cause," and no question of a mysterious non-local (super-luminal) interaction, changing a value on one system due to a measurement on the other system, can possibly arise.

variously associated with the chemical revolution. Kant never brought this projected work to completion; but we can here see, nonetheless, how the original Kantian (meta-)framework was already strained to the limit by the new turn-of-the-century scientific developments – developments which eventually led to the decisive overthrow of the Kantian (meta-)framework at the beginning of the next century.[64]

This brings me to my second point. The chemical revolution played a central role in the initial stages of what I referred to, in Lecture III, as the "second scientific revolution" (Part One, note 62 above). For it was here that quantitative accuracy and precision was first introduced into experimental practice in a wide variety of related fields, including chemistry (along with the theory of gases), the theory of heat, electricity, and magnetism. And it was precisely these developments, moreover, that led to the formulation of radically new styles of theorizing in mathematical physics – including the wave theory of light, thermodynamics and statistical mechanics, and James Clerk Maxwell's theory of the electro-magnetic field – whereby the disciplines Kuhn has called "Baconian" sciences were finally incorporated within the type of mathematical theorizing long associated with the "classical" sciences.[65] But the most important point, in our present context, is that it was also precisely these developments which comprised the indispensable background to the two great revolutions of twentieth century mathematical physics: relativity and quantum mechanics. Indeed, both twentieth century revolutions arose directly from late nineteenth century attempts to comprehend the interaction of light (more generally, electro-magnetic radiation) with matter, as this interaction was used to probe the new atomic perspective on the microstructure of matter first opened up in the chemical revolution.[66] The chemical revolution – and, more generally, the "second scientific revolution" of which it was an integral part – is therefore intimately intertwined with our

---

64. For discussion of Kant's late work (collected together in what we now know as the *Opus postumum*) on the *Transition* project and the chemical revolution see Friedman (1992, chapter 5).

65. See again Kuhn (1976/77).

66. The indispensable background to relativity is Lorentz's theory of the electron, which, among other things, explains the interaction between electro-magnetic radiation and matter in terms of resonant frequencies of electrons bound in an atom resulting from (as well as contributing to) such interaction. The indispensable background to quantum mechanics is of course the original Bohr model of the atom, with its explanation of atomic spectra. And it is quantum mechanics as well that completes the chemical revolution, by giving a physical explanation of both the periodic table of elements and electronic valence. All of these modern contributions to atomic theory have their roots in John Dalton's early nineteenth century chemical atomism, which further exploits Lavoisier's use of the balance via the fundamental law of combining proportions.

present account at both ends: with Kant's attempt to extend the reach of his own philosophical (meta-)framework to the emerging new sciences at the turn of the nineteenth century, and with the great conceptual revolutions of the early twentieth century that conclusively signalled the downfall of Kant's original (meta-)framework and the consequent need for its fundamental modification.

The Darwinian Revolution. These far-reaching developments in late nineteenth century biology have even less of a direct connection with our present philosophical framework. The primary innovations in this case were not mathematical in nature, and so there is no question at all here of coordinating or (more generally) constitutive principles in our sense. Nevertheless, as in the case of the chemical revolution, there are still indirect connections with some of the most important events highlighted in the present account, both within the sciences and at the philosophical or meta-scientific level. And these connections, I believe, shed further light on our present philosophical predicament vis-à-vis the sciences.

In the first place, Kant's attempt, in the *Critique of Judgement,* to explain how the physical sciences and the life sciences can be comprehended within a common philosophical (meta-)framework was an important part of the background to the late nineteenth century revolution in biology. In particular, Kant's attempt to show, in his terms, how mechanism and teleology can be consistently fitted together within a single unified picture of the natural world supplied an important model for a non-reductionist but also non-vitalist approach to the relationship between biology and physics. Accordingly, Kant's work in the *Critique of Judgement* was a central part of the background for the assimilation of the emerging new evolutionary ideas, especially in Germany where a teleological understanding of evolution was particularly prevalent.[67] Since, as we have just seen, Kant's work in the *Critique of Judgement* also served as a focus for his attempt to comprehend the emerging new ideas in the physical sciences as well (where, in particular, the relationship between what Kant calls constitutive and regulative principles became a crucially important issue),[68] it also provided a turn-of-the-century focal point for the relationship between these ideas and the life sciences as well.

In the second place, the question of the relationship between the physical

---

67. For discussion of this German context see Lenoir (1982).

68. For discussion of this issue in the context of the *Opus postumum* see Friedman (1992, chapter 5, section II). Compare also Part One, note 77 above.

and the life sciences played a key role in some of the nineteenth century developments we have already reviewed constituting the indispensable background to the development of Einsteinian relativity theory. Helmholtz's work in the foundations of geometry, in particular, was framed within his own contributions to the emerging new science of psycho-physics – a science which, as its very name suggests, was explicitly intended to bridge the traditional philosophical gulf between matter and the mind. In particular, Helmholtz took his neo-Kantian perspective on space perception and the principle of causality (the lawfulness of nature) to show that and how the mind (and more generally life) can be fully integrated within the picture of the world due to mathematical physics and, simultaneously, as supplying the necessary or "transcendental" constitutive framework for this very same physical world-picture. More generally, all the great contributors to nineteenth century psycho-physics – including, besides Helmholtz, Johannes Müller, Gustav Fechner, Ewald Hering, and Ernst Mach – understood themselves to be standing on the threshold of a new intellectual era in which physics, psychology, and biology were finally to be united within a single rigorous scientific framework.

In the work of Hering and Mach, however, the theory of space perception took a more explicitly evolutionary turn that was fundamentally opposed to Helmholtz's original conception. For Helmholtz, our representation of space is fully learned or acquired by what he conceived of as an *individual* adaptation, in that, according to Helmholtz's "empiricist" theory of perception, each individual acquires the representation of space within its own lifetime, and there is absolutely no inheritance, in this regard, from previous generations. But Hering and Mach explicitly defended the opposing "nativist" position, according to which the representation of space is largely "wired-in" to individual physiology at birth and is therefore subject to a truly evolutionary adaptation extending across many generations.[69] And it was precisely this pycho-physiological position, in turn, that formed the indispensable background for Mach's contrasting, less Kantian and more pragmatically naturalistic conception of scientific epistemology, according to which all of the sciences (including the mathematical and physical sciences) are constituents of a fundamentally biological evolutionary process aiming at a holistic

---

69. For discussion of this nineteenth century psycho-physiological debate between "empiricism" and "nativism" see Turner (1994). As Hatfield (1990, Appendix A) is careful to point out, this debate should not be confused with the traditional philosophical debate between "empiricism" and "rationalism." Indeed, from a philosophical point of view, it is clearly Helmholtz who is more of a "rationalist," Mach who is more of an "empiricist."

adaptation of the human race to its natural environment.[70] Moreover, a closely related epistemological perspective also framed the development of American pragmatism – as is particularly evident in the work of Charles Sanders Peirce, where a characteristically late nineteenth century view of both nature and science, as a part of nature, developing according to (probabilistic or "tychistic") evolutionary processes, is searchingly and imaginatively explored.[71]

Yet the further development of the physical sciences in the early twentieth century by no means conformed to this late nineteenth century "pan-biological" vision. We instead saw the creation of extraordinarily abstract mathematical theories (relativity and quantum mechanics), self-consciously designed to move further and further from any natural connection at all with human intuition or ordinary sensory experience. Indeed, in light of our discussion at the end of section 4 above, the fundamental problem here can be identified rather precisely. The entire late nineteenth century discussion of space perception, against the background of contemporaneous developments in biology and psycho-physics, is explicitly framed within the Helmholtzian theory of free mobility. But this theory, of course, is limited to spaces of constant curvature and is thus quite inapplicable to the radically new (space-time) structure employed in the general theory of relativity. The characteristically late nineteenth century attempt to unify biology and psycho-physiology with the underlying mathematical framework of fundamental physical theory therefore breaks down at this point, and we are left instead with a new, characteristically twentieth century conception of physical theorizing that aligns it with modern abstract mathematics – and even with modern mathematical logic – rather than with any concurrent developments within the life sciences.

Finally, whereas the early twentieth century scientific philosophizing of the logical empiricists was motivated and sustained by precisely this new (abstractly mathematical) conception of physical theorizing, the parallel scientific philosophizing of the American pragmatist movement continued to place its primary emphasis on the life sciences – and, more particularly, on

---

70. Here I am especially indebted to discussions with Paul Pojman, whose dissertation on this topic is Pojman (2000). As Pojman points out, both Hering and Mach operated within the more explicitly teleological German understanding of evolution and, in addition were both Lamarckians. For Mach, this made cultural and historical evolution continuous with biological evolution. I am also indebted to Michael Heidelberger and Rasmus Winther for discussion of these matters.
71. This viewpoint is most explicitly developed in a series of five papers originally published in *The Monist* in 1891–92: see Hartshorne and Weiss (1931-35, vol. 6, §§ 7–65, 102–63, 238–71, 287–317). I am indebted to Elisabeth Lloyd, and also to Peter Godfrey-Smith, for discussions of the relationship between evolutionary biology and American pragmatism.

Darwinian evolutionary biology. The failure of the late nineteenth century ambition to comprehend both the life sciences and the physical sciences by means of a common scientific paradigm (which was to be fundamentally evolutionary in nature) and a common philosophical meta-framework (some or another version of evolutionary pragmatic naturalism) has thus given rise to a contemporary split within scientific philosophy between those tendencies arising from the logical empiricist tradition and those inspired by American pragmatism. But further discussion of this idea must definitely be left to another occasion.

References

Barnes, B. (1982) *T. S. Kuhn and Social Science*. London: Macmillan.

Barnes, B. and D. Bloor (1982) "Relativism, Rationalism and the Sociology of Knowledge." In M. Hollis and S. Lukes, eds. *Rationality and Relativism*. Oxford: Blackwell.

Bellone, E. (1976) *Il mondo di carta: Ricerche sulla seconda rivoluzione scientifica*. Milan: Mondadori. Translated as *A World on Paper: Studies on the Second Scientific Revolution*. Cambridge, Mass.: MIT Press, 1980.

Benacerraf, P. and H. Putnam, eds. (1964) *Philosophy of Mathematics: Selected Readings*. Englewood Cliffs, N.J.: Prentice-Hall. Second edition, Cambridge: Cambridge University Press, 1983.

Birkhoff, G. and J. von Neumann (1936) "The Logic of Quantum Mechanics." *Annals of Mathematics* 37, 823–843.

Bohr, N. (1939) "The Causality Problem in Atomic Physics" (with following discussion). In *New Theories in Physics*. Paris: International Institute of Intellectual Co-operation.

Bricker, P. and R. Hughes, eds. (1990) *Philosophical Perspectives on Newtonian Science*. Cambridge, Mass.: MIT Press.

Brittan, G. (1978) *Kant's Theory of Science*. Princeton: Princeton University Press.

Carnap, R. (1922) *Der Raum: Ein Beitrag zur Wissenschaftslehre*. Berlin: Reuther and Reichard.

———. (1928) *Der logische Aufbau der Welt*. Berlin: Weltkreis. Translated as *The Logical Structure of the World*. Berkeley and Los Angeles: University of California Press, 1967.

———. (1934) *Logische Syntax der Sprache*. Wien: Springer. Translated as *The Logical Syntax of Language*. London: Kegan Paul, 1937.

———. (1950) "Empiricism, Semantics, and Ontology." *Revue Internationale de Philosophie* 11, 20–40. Reprinted in *Meaning and Necessity*. Second edition. Chicago: University of Chicago Press, 1956.

———. (1963) "Replies and Systematic Expositions." In Schilpp (1963).

Cassirer, E. (1910) *Substanzbegriff und Funktionsbegriff: Untersuchungen über die Grundfragen der Erkenntniskritik*. Berlin: Bruno Cassirer. Translated as *Substance and Function*. Chicago: Open Court, 1923.

Cohen, I. (1999) "A Guide to Newton's *Principia*." In I. Cohen and A. Whitman, eds. *Isaac Newton: The Principia*. Berkeley and Los Angeles: University of California Press.

Creath, R., ed. (1990) *Dear Carnap, Dear Van: The Quine-Carnap Correspondence and Related Work*. Berkeley and Los Angeles: University of California Press.

De Pierris, G. (1993) "The Constitutive A Priori." *Canadian Journal of Philosophy*, Supplementary Volume 18, 179–214.

DeWitt, B. and N. Graham, eds. (1973) *The Many-Worlds Interpretation of Quantum Mechanics*. Princeton: Princeton University Press.

DiSalle, R. (1990) "The 'Essential Properties' of Matter, Space, and Time." In Bricker and Hughes (1990).

———. (1991) "Conventionalism and the Origins of the Inertial Frame Concept." *PSA 1990*, vol. 2, 139–147.

———. (1995) "Spacetime Theory as Physical Geometry." *Erkenntnis* 42, 317–337.

Earman, J. (1993) "Carnap, Kuhn, and the Philosophy of Scientific Methodology." In Horwich (1993).

Earman, J. and M. Friedman (1973) "The Meaning and Status of Newton's Law of Inertia and the Nature of Gravitational Forces." *Philosophy of Science* 40, 329–359.

Edgerton, S. (1991) *The Heritage of Giotto's Geometry: Art and Science on the Eve of the Scientific Revolution*. Ithaca: Cornell University Press.

Ehlers, J. (1981) "Über den Newtonschen Grenzwert der Einsteinschen Gravitationstheorie." In J. Nitsch, J. Pfarr, and E. Stachow, eds. *Grundlagenprobleme der modernen Physik*. Mannheim: Bibliographisches Institut.

Ehlers, J., F. Pirani, and A. Schild (1972) "The Geometry of Free Fall and Light Propagation." In L. O'Raifeartaigh, ed. *General Relativity. Papers in Honour of J. L. Synge*. Oxford: Oxford University Press.

Einstein, A. (1916) "Die Grundlage der allgemeinen Relativitätstheorie." *Annalen der Physik* 49, 769–822. Translated as "The Foundation of The General Theory of Relativity." In H. Lorentz, et. al., *The Principle of Relativity*. London: Methuen, 1923.

———. (1917) *Über die spezielle und die allgemeine Relativitätstheorie, gemeinverständlich*. Braunschweig: Vieweg. Translated as *Relativity, the Special and the General Theory: A Popular Exposition*. London: Methuen, 1920.

———. (1921) "Geometrie und Erfahrung." *Preussische Akademie der Wissenschaft. Physikalisch-mathematische Klasse. Sitzungsberichte*, 123–130. *Erweiterte Fassung des Festvortrages gehalten an der Preussischen Akademie der Wissenschaft zu Berlin am 27. Januar 1921*. Berlin: Springer. Translated as "Geometry and Experience." In G. Jeffrey and W. Perrett, eds. *Sidelights on Relativity*. London: Methuen, 1923.

———. (1922) *The Meaning of Relativity*. Princeton: Princeton University Press.

Friedman, M. (1983) *Foundations of Space-Time Theories: Relativistic Physics and the Philosophy of Science*. Princeton: Princeton University Press.

———. (1990) "Kant and Newton: Why Gravity is Essential to Matter." In Bricker and Hughes (1990).

———. (1992) *Kant and the Exact Sciences*. Cambridge, Mass.: Harvard University Press.

———. (1996) "Exorcising the Philosophical Tradition: Comments on John McDowell's *Mind and World*." *Philosophical Review* 105, 427–467.

———. (1997a) "Philosophical Naturalism," *Proceeding and Addresses of The American Philosophical Association* 71, 7–21.

———. (1997b) "Helmholtz's *Zeichentheorie* and Schlick's *Allgemeine Erkenntnislehre*: Early Logical Empiricism and Its Nineteenth-Century Background." *Philosophical Topics* 25, 19–50.

————. (1998) "On the Sociology of Scientific Knowledge and its Philosophical Agenda." *Studies in History and Philosophy of Science* 29, 239–271.

————. (1999) *Reconsidering Logical Positivism*. Cambridge: Cambridge University Press.

————. (2000a) "Geometry, Construction, and Intuition in Kant and His Successors." In G. Scher and R. Tieszen, eds. *Between Logic and Intuition: Essays in Honor of Charles Parsons*. Cambridge: Cambridge University Press.

————. (2000b) *A Parting of the Ways: Carnap, Cassirer, and Heidegger*. Chicago: Open Court.

————. (2001) "Geometry as a Branch of Physics: Background and Context for Einstein's 'Geometry and Experience.'" In D. Malament, ed. *Reading Natural Philosophy: Essays in the History and Philosophy of Science and Mathematics to Honor Howard Stein on His 70th Birthday*. Chicago: Open Court.

Goodman, N. and W. Quine (1947) "Steps Toward a Constructive Nominalism." *Journal of Symbolic Logic* 12, 97–122.

Guicciardini, N. (1999) *Reading the Principia: The Debate on Newton's Mathematical Methods for Natural Philosophy from 1687 to 1736*. Cambridge: Cambridge University Press.

Habermas, J. (1981) *Theorie des Kommunikativen Handelns*. 2 vols. Frankfurt: Suhrkamp. Translated as *The Theory of Communicative Action*. Boston: Beacon, 1984.

Hartshorne C. and P. Weiss, eds. (1931–35) *Collected Papers of Charles Sanders Peirce*. 6 vols. Cambridge, Mass.: Harvard University Press.

Hatfield, G. (1990) *The Natural and the Normative: Theories of Spatial Perception from Kant to Helmholtz*. Cambridge, Mass.: MIT Press.

Heisenberg, W. (1958) *Physics and Philosophy: The Revolution in Modern Science*. New York: Harper & Row.

Helmholtz, H. (1865) *Vorträge und Reden*. 2 vols. Braunschweig: Vieweg. Fifth Edition, 1903.

Hempel, C. (2000) *Selected Philosophical Essays*, ed. R. Jeffrey. Cambridge: Cambridge University Press.

Hertz, P. and M. Schlick, eds. (1921) *Hermann v. Helmholtz: Schriften zur Erkenntnistheorie*. Berlin: Springer. Translated as *Hermann von Helmholtz: Epistemological Writings*. Dordrecht: Reidel, 1977.

Horwich, P., ed. (1993) *World Changes: Thomas Kuhn and the Nature of Science*. Cambridge, Mass.: MIT Press.

Howard, D. and J. Stachel, eds. (1989) *Einstein and the History of General Relativity*. Boston: Birkhäuser.

Köhnke, K. (1986) *Entstehung und Aufstieg des Neukantianismus: die deutsche Universitätsphilosophie zwischen Idealismus und Positivismus*. Frankfurt: Suhrkamp. Translated (partially) as *The Rise of Neo-Kantianism*. Cambridge: Cambridge University Press, 1991.

Kuhn, T. (1959) "Energy Conservation as an Example of Simultaneous Discovery." In M. Clagett, ed. *Critical Problems in the History of Science*. Madison: University of Wisconsin Press, 1959. Reprinted in Kuhn (1977).

————. (1962) *The Structure of Scientific Revolutions*. Chicago: University of Chicago Press. Second edition, 1970.

————. (1973) "Objectivity, Value Judgment, and Theory Choice." Matchette Lecture, Furman University. Reprinted in Kuhn (1977).

————. (1976) "Mathematical versus Experimental Traditions in the Development of Physical Science." *Journal of Interdisciplinary History* 7, 1–31. Reprinted in Kuhn (1977).

————. (1977) *The Essential Tension: Selected Studies in Scientific Tradition and Change.* Chicago: University of Chicago Press.

————. (1993) "Afterwords." In Horwich (1993).

Lenoir, T. (1982) *The Strategy of Life: Teleology and Mechanism in Nineteenth Century German Biology.* Dordrecht: Reidel.

Lewis, C. (1929) *Mind and the World Order.* New York: Scribner.

Lützen, J. (1995) "Interactions between Mechanics and Differential Geometry in the 19th Century." *Archive for History of Exact Sciences* 49, 1–72.

Malament, D. (1986) "Newtonian Gravity, Limits, and the Geometry of Space" In R. Colodny, ed. *From Quarks to Quasars: Philosophical Problems of Modern Physics.* Pittsburgh: University of Pittsburgh Press.

McDowell, J. (1994) *Mind and World.* Cambridge, Mass.: Harvard University Press.

Miller, A. (1981) *Albert Einstein's Special Theory of Relativity.* Reading, Mass.: Addison-Wesley.

Misner, C., K. Thorne, and J. Wheeler (1973) *Gravitation.* San Francisco: Freeman.

Mulder, H. and B. van de Velde-Schlick, eds. (1978) *Moritz Schlick: Philosophical Papers.* Vol. 1. Dordrecht: Reidel.

Norton, J. (1984) "How Einstein Found His Field Equations, 1912–1915." *Historical Studies in The Physical Sciences* 14, 253–316. Reprinted in Howard and Stachel (1989).

————. (1985) "What Was Einstein's Principle of Equivalence?" *Studies in History and Philosophy of Science* 16, 203–246. Reprinted in Howard and Stachel (1989).

Pap, A. (1946) *The A Priori in Physical Theory.* New York: Columbia University Press.

Parrini, P. (1998) *Knowledge and Reality.* Dordrecht: Kluwer.

Poincaré, H. (1902) *La Science et l'Hypothèse.* Paris: Flammarion. Translated as *Science and Hypothesis.* In G. Halsted, ed. *The Foundations of Science.* Lancaster: Science Press, 1913.

————. (1905) *La Valeur de la Science.* Paris: Flammarion. Translated as *The Value of Science.* In G. Halsted, *op. cit.*

————. (1908) *Science et Méthode.* Paris: Flammarion. Translated as *Science and Method.* In G. Halsted, *op. cit.*

Pojman, P. (2000) *Ernst Mach's Biological Theory of Knowledge.* Doctoral Dissertation: Indiana University.

Quine, W. (1948) "On What There Is." *Review of Metaphysics* 2, 21–38. Reprinted in Quine (1953).

————. (1951) "Two Dogmas of Empiricism." *Philosophical Review* 60, 20–43. Reprinted in Quine (1953).

————. (1953) *From a Logical Point of View.* New York: Harper.

————. (1955) "The Scope and Language of Science." In L. Leary, ed. *The Unity of Knowledge.* New York: Doubleday. Reprinted in *The Ways of Paradox and Other Essays.* New York: Random House, 1966.

————. (1960) *Word and Object.* Cambridge, Mass.: MIT Press.

————. (1970) *Philosophy of Logic.* Englewood Cliffs, N.J.: Prentice-Hall.

Rawls, J. (1993) *Political Liberalism.* New York: Columbia University Press. Second edition, 1996.

Reichenbach, H. (1920) *Relativitätstheorie und Erkenntnis Apriori.* Berlin: Springer. Translated as *The Theory of Relativity and A Priori Knowledge.* Los Angeles: University of California Press, 1965.

Reisch, G. (1991) "Did Kuhn Kill Logical Empiricism?" *Philosophy of Science* 58, 264–277.

Richards, J. (1977) "The Evolution of Empiricism: Hermann von Helmholtz and the Foundations of Geometry." *British Journal for the Philosophy of Science* 28, 235–253.

Riemann, B. (1919) *Über die Hypothesen, welche der Geometrie zugrunde liegen. Neu herausgegeben und erläutert von H. Weyl.* Berlin: Springer.

Schilpp, P., ed. (1949) *Albert Einstein: Philosopher-Scientist.* La Salle: Open Court.

———., ed. (1963) *The Philosophy of Rudolf Carnap.* La Salle: Open Court.

Schlick, M. (1917) *Raum und Zeit in der gegenwärtigen Physik.* Berlin: Springer. Translated as *Space and Time in Contemporary Physics.* In Mulder and van de Velde Schlick (1978).

———. (1918) *Allgemeine Erkenntnislehre.* Berlin: Springer. Translated as *General Theory of Knowledge.* La Salle: Open Court, 1985.

———. (1922) "Helmholtz als Erkenntnistheoretiker." In *Helmholtz als Physiker, Physiologe und Philosoph.* Karlsruhe: Müllersche Hofbuchhandlung. Translated as "Helmholtz the Epistemologist." In Mulder and van de Velde-Schlick (1978).

Sellars, W. (1956) "Empiricism and the Philosophy of Mind." In H. Feigl and M. Scriven, eds. *Minnesota Studies in the Philosophy of Science.* Vol. 1. Minneapolis: University of Minnesota Press.

Shapin S. and S. Schaffer (1985) *Leviathan and the Air-Pump: Hobbes, Boyle, and the Experimental Life.* Princeton: Princeton University Press.

Stachel, J. (1980) "Einstein and the Rigidly Rotating Disk." In A. Held, ed., *General Relativity and Gravitation.* New York: Plenum. Reprinted as "The Rigidly Rotating Disk as the 'Missing Link' in the History of General Relativity." In Howard and Stachel (1989).

Stein, H. (1967) "Newtonian Space-Time." *Texas Quarterly* 10, 174–200.

———. (1977) "Some Philosophical Prehistory of General Relativity." In J. Earman, C. Glymour, and J. Stachel, eds. *Minnesota Studies in the Philosophy of Science.* Vol. VIII. Minneapolis: University of Minnesota Press.

Torretti, R. (1978) *Philosophy of Geometry from Riemann to Poincaré.* Dordrecht: Reidel.

———. (1983) *Relativity and Geometry.* New York: Pergamon.

Turner, R. (1994) *In the Eye's Mind: Vision and the Helmholtz-Hering Controversy.* Princeton: Princeton University Press.

van Fraassen, B. (1980) *The Scientific Image.* Oxford: Oxford University Press.

Westfall, R. (1971) *Force in Newton's Physics: The Science of Dynamics in the Seventeenth Century.* New York: American Elsevier.

Wigner, E. (1961) "Remarks on the Mind-Body Question." In I. Good, ed. *The Scientist Speculates.* London: Heinemann. Reprinted in *Symmetries and Reflections.* Bloomington: Indiana University Press, 1966.

Wittgenstein, L. (1922) *Tractatus Logico-Philosophicus.* London: Routledge.

Index